MW01181199

Cover Art

Juan Martin Alfaro, MFA

02/21/06

DEAR ED:

I WANT TO THANK YOO VERY MUCH
FOR ALL YOUR HELP.
YOUR GREAT CONTRIBUTION MADE THIS
POSSIBLE.

AFFECTIONATELY

Juan

ISBN: 0-9772637-0-3

Published by Drylongso Publications
2001 M Street, NW, Box 19795, Washington, DC 20036-9795

Water For All in Latin America

A daunting task

Juan Alfaro, B.S., M.S., P.E.

Dedicated to

Olga María, my beloved wife, my children and my grands for their support and inspiration.

Contents

Glossary & Abbreviations

AID	US Agency for International Development
AIDIS	Inter American Association of Sanitary Engineering and Environmental Sciences
BOD_5	Biochemical Oxygen Demand
BOOT	Build-Own-Operate-Transfer
BOT	Build-Operate-Transfer
CAAD	Computer Aid for Design and Drawing
CEMAP	Centro Mesoamericano de Estudios Sobre Tecnologia Apropiada
CEPIS	Centro Panamericano de Ingenieria Sanitaria y Ciencias del Ambiente
COD	Chemical Oxgen Demand
CONSUMPTION	Water demand at zero price
CVM	Contingency Valuation Methodology
DANIDA	Danish International Development Agency
DBP	Disinfection by Products
DEMAND	Water comsumption at different prices
ECO	Pan American Center for Human Ecology and Health (Mexico)
EIA	Environmental Impact Assessment
EIRR	Economical Internal Rate of Return
GTZ	German Agency for Technical Cooperation
IADB	Inter American Development Bank
IDRC	Canadian International Development-Research Center
IIG/JICA	Japanese International Cooperation Agency
IRR	Internal Rate of Return
JTU, NTU	Turbidity units
Lppd	Liters per person per day
Lt	Liter

Lts/sec	Liters per second
m,mt	Meter
ml	Milliliter
MPN	Most Probable Number (coliform Index)
mt/sec	meter per second
NGO	Non-Governmental Organizations
NH_3	Ammoniacal Nitrogen
NPV	Net Present Value
O&M	Operation and Maintenance
ONE BILLION	1,000 million
p.p.m.	Milligrams per liter
PAHO	Pan American Health Organization
PPP	Public Private Partnership
PSP	Private Sector Participation
SS	Suspended Solids
UFW	Unaccounted For Water
UN	United Nations
UNDP	United Nations Development Program
US $	United States dollar
USEPA	U.S. Environmental Protection Agency
WB	World Bank
WHO	World Health Organization
WHO/UNEP	World Health Organization/United Nations Environmental Program
WTP	Willigness to Pay
WWTP	Waste Water Treatment Plant

Preface

Water for All, might be an axiom in the water field. Providing *Water for All* involves the consideration of many variables including the achievement of the people's well being through the increase of water and wastewater service coverage (health factor), the people's willingness to pay (economic factor), the people's capability to pay (financial factor), maintaining environmental quality, and minimization of cost. A primary consideration is finding the proper institutional approach to serve the poor located in the scattered areas of the large cities (Peri-Urban), and the poor located in the rural areas of Latin America. Furthermore, it is important to note that the water field has not been receiving priority in national plans due to the economic burden on the governments' resources having been largely committed to combating poverty as a first priority and recently to fight the corruption and terrorism. Traditionally, water has been considered a gift to all from mother nature. Furthermore, people feel it should be free – an assumed "right." But, consistently safe, reliable drinking water, in sustainable quality and quantity to meet the demand implies a cost that someone has to pay. The criterion of quality and quantity sustainability is very important and indispensable to a satisfactory system.

The acceptance of the concept "price elasticity of water demand," requires a market analysis in order to supply drinking water under economic (willingness to pay) and environmental restrictions, which have not been consistently considered in the past.

The author, currently President of CEP International Inc, a consulting firm based in the Washington, DC area, acquired his expertise in the water field first in his country of origin, Peru, second as Head of the Water Sector (Water, Wastewater, Solid Waste and Air) at the Inter American Development Bank (IADB), in Washington DC, for a period of 20 years; third, as an international consultant for the past 10 years working on, among other areas, on institutional development, including private sector participation (PSP) for water utilities in Latin America. This book is a summary of the author's experience in preparing and appraising projects and it is complementary to his previous book, in Spanish, titled: "Tres Decadas de Saneamiento en Latinoamerica," published in 1999[1]. The author acknowledges the review and suggestions for this volume from Dr. Edward Martin (Director of CEP's Environmental Division); Mr. Alnunfo Carrandi (Director of CEP's Institutional and Financial Division); Dr. Jorge Ducci (CEP's Principal associate of the Economic Division).

Finally, in order to address properly the nature of the "water for all" problem, it is necessary to perform a comprehensive analysis of all the variables and alternatives involved in the design of a sound water and/or wastewater investment project, whether it is financed locally or internationally, with or without private sector participation. In addition to the technical issues, environmental quality, institutional models, private sector participation, and financial, economic and political factors, are discussed as well.

PURPOSE

The purpose of this book is to present in a comprehensive manner all the variables related to design, evaluation, implementation and sustainability of investment projects in the water and wastewater field in Latin America.

It is recognized that there are many publications about the subject prepared by international organizations and individuals. It is not the intention of this book to pretend to substitute for them, but rather to complement them by presenting in a consolidated form the several alternatives and options that should be listed, studied and compared in order to arrive at the optimum approach for a particular case. In some way, the book might be considered as a checklist for project preparation.

OUTLINE

The book contains seven chapters that might be consulted separately or as a whole, according to the need of the reader. Chapter One discusses the evolution of the water field in Latin America up to about the year 2000. Problems and deficiencies are also included as well as a summary of costs necessary to alleviate the situation at the regional level, as recommended by the new Water Decade (2005-2015), scheduled to be launched by the United Nations. Chapter Two is a summary of the pertinent policies of the Inter American Development Bank (IADB). It is noted that project design must comply with policies established by the lending agencies as well as by local institutions. Policies are similar in all international agencies but it is worthwhile to review them. Chapter Three is a guideline of the criteria, procedures and methodologies to evaluate an investment project. It is divided in five sections as follows:

1. A checklist of the technical requirements that every project should contain. A comparison of alternatives to design water and wastewater projects for urban areas is discussed.
2. A description of the adopted methodology to measure environmental impacts and to determine and quantify mitigation measures for the negative impacts.
3. A description of the institutional schemes that might be adopted by the public utilities in Latin America to implement projects, including the trend toward private sector participation (PSP).
4. A brief description of the current methodologies adopted to demonstrate project financial feasibility.
5. A description of methodologies to determine project economic feasibility based on willingness to pay (WTP), developed directly from the results of household surveys.

Chapter Four is a description of the different policies and alternatives to provide safe water to the poor in urban areas. Rural area analysis is not included due to the large experience gained in the Latin American region and the prevalence for many years of the Co-Operatives model that has proven to be successful. Chapter Five is a brief review of the institutional models adopted by the public utilities in 12 selected countries of Latin America presented as a representative sample of the whole field, and some additional case studies. Chapter Six is a description of the current types of public sector participation. Chapter Seven is devoted to conclusions and recommendations.

Chapter One

Evolution of the Coverage of Services in Latin America

Evolution of the Coverage of Services in Latin America

The Status of the Field up to 1990

Traditionally, the development of the field of water and wastewater in Latin America has always been dependent on national resources, and the allocation of resources for the water supply area varies from country to country. In some countries like Argentina and Peru, having been very good exporters of raw materials in the past, ambitious national water plans were carried out in the early 1900's. But in other countries the situation was and is not the same. The information presented in this chapter comes from reference (2).

At the beginning of the 1960's decade, the coverage of water and wastewater services in Latin America (percentage of the population served) was very limited and the services inadequate. Out of an estimated urban population of 100 million, only about 40 million (40%) had in-house connections for water service and another 20 million (20%) were served solely by public standpipes or other mechanisms. About 30 million (30%) of the same urban population had wastewater connections to a public service without treatment.

In rural areas the situation was even worse since out of 100 million, barely 6% had easy access to water, and wastewater treatment and disposal was practically nonexistent except of course, for some rudimentary latrines. At that stage there were not national plans in most of the countries. Financing came from national budgets, the operation cost of the systems was subsidized, no community participation in the decision-making process existed, and the government authorities in charge did not apply efficient financial management.

The first concerted effort of Latin America's countries to improve the water and wastewater service sector is embodied in the Charter of Punta del Este (Uruguay), which was signed in late 1960. By means of that document the signature governments undertook the task to raise the coverage of water and wastewater service to 70% in the urban areas and 50% in the rural areas. The Alliance for Progress launched by the US government was instrumental in establishing for the first time, defined targets and also financially supporting the countries' efforts. The Interamerican Development Bank (IADB) created at the end of 1959, was the institution in charge of channeling the economic resources allocated by the USA to the water and wastewater sector. It is not a coincidence that the first IADB loan, for US$3.3 million, was approved in January 1961 to partially finance the water and wastewater Master Plan for the city of Arequipa, Peru, SA. The IADB was also the pioneer international institution to finance projects in the rural sector.

At the end of the decade (1961-1970), the population served with water in the urban areas had doubled and that in the rural areas had tripled. Out of 148 million urban inhabitants, 80 million (54%) had water in-house service, and in the rural area 20 million (16%) were receiving safe water. In the urban sector, about 51 million (34%) had wastewater service and in the rural sector out of 118 million, 11 million (9%) had some form of excreta disposal. Despite the substantial increase in the population served, the goals established by the Alliance for Progress were not achieved due primarily to the growth of urban population and greater urbanization[(3)].

At the beginning of the next decade (1971-1980), the Ministers of Health of the region, met in Santiago de Chile, Chile, and signed a Ten Year Health Plan which established new coverage goals: for urban water supply, 80%, for urban wastewater, 70%, and for water and excreta disposal in the rural areas, 50%. In addition, the Ministers emphasized the need to formulate policies and execute programs to control water quality, air and soil pollution that were compatible with basic environmental sanitation, industrialization and urbanization.

At the end of 1980, the status was as follows. In the urban area out of 217 million , about 163 million inhabitants (75%) had water service and about 100 million (46%) had wastewater service. In the rural area out of 126 million, 24 million (19%) had safe water and about 15 million (12%) had some system of excreta disposal. Again, the new targets for water and wastewater in urban and rural areas were not reached.

The World Water Conference held in 1977 in Mar de Plata, Argentina, agreed to designate the period 1981-1990 the "Drinking Water and Environmental Sanitation Decade" simply "The Water Decade"[(4)]. It was also agreed to reach the following targets. Firstly, the commitment of the governments to provide to the entire population, if possible by 1990, with appropriate water and wastewater services and to give priority to areas in which water was lacking, to the poor and less favored. Secondly, to assign a larger amount of funds and personnel to the environmental sanitation area so as to better serve the population. Due to the economic struggle suffered by Latin American countries in this decade, also named the "lost decade," the new targets were not achieved.

At the end of 1990, the status was as follows: in the urban area out of 291 million, about 230 million inhabitants (79%) had water and about 142 million inhabitants (49%) had wastewater service. In the rural area, out of 124 million, 48 million (38%) had water and 28 million inhabitants (22%) had some form of excreta disposal.

As representative of the problems commonly faced by the countries in the delivery of water and wastewater services to the urban area, the results of one case study conducted by the author are presented herein.

Case Study[(5)]. In the urban water sector a survey was conducted in 1986 by the author to detect the main problems and specifically, operation and maintenance problems of the systems. Responses from 19 countries were received and the inquiries were presented in two parts.

Part 1.

a. General characteristics of the entity, Descriptive information.
b. Functional organization chart of the entity.
c. Details of personnel assigned (type and number) to maintenance. Detailed information .to obtain ratios related to the total number of personnel.
d. Annual expenses for maintenance during the previous 5 years. Budget for O&M.
e. Annual expenses for equipment, vehicles, tools, instruments etc., for the past 5 years. Details of the expenditures.

Part 2.

a. Philosophy of the entity on maintenance; political will.

b. Human resources allocated to maintenance
c. Organization of maintenance department
d. Allocation of financial resources for maintenance. Annual Budget.
e. Training programs.
f. Water quality

The adopted scale considered only 3 indicators: Deficient, Good, and Very Good. The results obtained were:

a. Philosophy: Deficient
b. Human resources: Deficient
c. Institutional organization: Good
d. Annual budget: Deficient
e. Training: Deficient
f. Water quality: Deficient

Almost all of the six factors were evaluated as "Deficient." Furthermore, personal experience indicates that the systems in Latin America suffer chronic deficiency in operation and maintenance (O&M). Hence, the conclusion was that rehabilitation of the current infrastructure, and the application of "best practices" for maintenance should be implemented universally. It is commonly agreed that the first task of a new management should be to concentrate on O&M. The level of unaccounted for water (UFW) in Latin America is a very significant problem and is a symptom of poor O&M practice. In most of the systems the UFW value is over 50% of the water produced.

PAHO's Regional Report 2000

The Pan American Health Organization (PAHO), as part of the World Health Organization (WHO) has consistently bolstered the development of the field over the past 50 years, and recently published the Regional Report on the Evaluation 2000[(6)]. The report contains updated and useful information about the water and wastewater sector. As shown in the report, for a total urban and rural population of 497 million, the status is as follows:

1. about 77 million inhabitants, do not have access to drinking water (house connections and/or easy access),
2. about 103 million inhabitants do not have some degree of sanitation (include house connections and in situ disposal), and,
3. about 208 million inhabitants have wastewater collection systems without treatment.

The above numbers clearly indicate that in order to continue reducing the gap in services provided, it is necessary to embark on a significant region-wide investment program.

Recently, the UN, based on the Regional Report, suggested launching a new Water Decade (period 2005-2015), in order to reduce the shortfall in water supply as well as wastewater service by 50%. The Regional Report estimates about US$100 billion are required to reach the recommended service coverage.

Based on past experience and the economic struggle faced nowadays by many countries in Latin America, it seems very difficult to invest a yearly average of US$10 billion, unless an innovative financial and management approach is envisioned. It should be noted that this amount is twice the maximum level expended yearly during the past decade. The hope of the New Decade is the mobilization of human and economic resources, technical consultants and system designers, construction contractors, operators, suppliers, and banks to arrive at significantly improved service coverage.

Summary of Water and Wastewater Service Status

Based on personal experience and findings of the Regional Report 2000, increasing delivery of the water and wastewater services, in compliance with Agenda 21 of the Conference on Environment, Río de Janeiro, Brazil, in 1992, possesses the bottleneck areas discussed below.

1. *Institutions.* Countries must establish a legislative framework sufficient to properly define and separate the roles of planning, operation and regulation. Strong regulatory agencies should be created with adequate financial and human resources. It is likely that new institutional models, different from those that exist presently, should be adopted.
2. *Political Will and Priority Setting.* The governments have to endorse the new Water Decade and work on the preparatory activities. A high priority should be assigned by the countries to the water and wastewater areas in their national plans. The author worked on the implementation of the last Water Decade and knows by experience the difficulties that have to be overcome in the countries.
3. *Decentralization.* Water and wastewater systems should be operated in a decentralized manner (states, departments, provinces, municipalities and rural co-operatives); more public participation and awareness of the problems should be

encouraged. Implementation is easier and cost recovery is more likely with the communities' concurrence.

4. *Financing.* Financial resources are indispensable to carry out programs. Public funding, even with international funding participation, has not been successful. More private sector participation, especially private sector financing, should be encouraged. Chile is a good example. The water and wastewater sector in that country is, by law, in private hands with limited state participation but with strong regulations. The US$ 100 billion recommended by PAHO, to reduce the deficit of services to 50% (period 2005- 2015), is a serious challenge for all the countries of the region.
5. *Rehabilitation and Maintenance.* Programs to rehabilitate the current infrastructure and to improve maintenance practices should be encouraged. Reduction of unaccounted for water is essential. Reduction to one-half the current value would increase the service coverage by a like percentage with the same infrastructure. Proper instrumentation and communication tools are of the first importance in order to improve control and operation of the existing systems
6. *Technology.* Use of imported technology has proven inefficient, difficult to operate, and prohibitively costly. Use of innovative, low cost technology (appropriate technology) should be encouraged. The application of low cost and simplified systems is of particular importance to serve the poor located in the peri-urban (scattered) areas of the cities.
7. *Human Resources.* In the Latin American region, there is concern over the training of water and wastewater utilities' staff responsible for the delivery of services. The creation of a system of training, perhaps through academia should be encouraged, for professionals and technicians in all aspects of sanitary end environmental engineering. Universities should be encouraged and technological centers should be developed, to promote research and do training. The status of human resources differs from country to country. The Centro Panamericano de Ingeniería

Sanitaria y Ciencias del Ambiente (CEPIS) is a good example to follow.

8. *Sector Coordination.* Location of the water and wastewater sector within the organizational structure of the State varies from country to country. Ministry of Health, Ministry of Urban Development, Ministry of the Presidency, are some examples of locations. There is a natural tendency to locate it within the health sector because water and wastewater services are considered the first step in the health staircase. At the launching of the Water Decade in 1980, Dr. H. Mahler, Director of WHO, quoted that in the future, "the number of house connections for 1000 people would be a better indicator of health than the number of beds at hospitals." The reason for the diversity of locations of the water and wastewater sector is due to the fact that water as an essential natural resource has multiple uses including agricultural and industrial. This consideration has led some countries to create national, state or departmental authorities to manage the resource in a holistic fashion giving priority to drinking water. However, holistic management often results in competition among water users, each of which have vastly different quality and quantity requirements.

Chapter Two

Review of the IADB's Sector Policies

Review of the IADB's Sector Policies

The Interamerican Development Bank's water and wastewater policies are summarized herein, noting that almost all the international lending banks and agencies have similar criteria. In essence, all of them want the proper allocation of their economic resources on sound and sustainable investment projects. Furthermore, it is noted that IADB has pioneered in some of these policies.

Basic Environmental Sanitation Policy

IADB approved this policy in 1984. The author is proud to have been instrumental in the preparation and approval of this policy(7). At that time, this policy responded to the current status of the field as well as added new criteria and considerations resulting from its 25 years of experience. The principal criteria are discussed below:

1. In all water and wastewater expansion projects, priority should be given to the rehabilitation of the existing infrastructure as a condition for further financing, and to optimize investments.
2. In every project, a sound local utility must exist to execute the project. This aspect of the policy was considered a stringent condition limiting the financing in most of the countries due to their institutional deficiencies early in the program. Hence, IADB's resources for the build up of the utility (through technical cooperation) should be included, if necessary, as a part of the loan. The disbursements should be tailored to the partial fulfillment of an institutional development plan.
3. Community participation is considered crucial for the success of projects. Sizing of the projects should be based on the interest of the

community and its participation in the different project activities, construction phase, partial financing, rates/tariffs, operation and maintenance.

4. Service levels should be based on the willingness to pay of the users. Appropriate low cost technologies should be chosen accordingly.
5. To design multiple urban water and wastewater programs, a selection procedure should be used. The selection criteria to consider are:
   a. location of the project in relation to the national development plan,
   b. population concentration,
   c. availability of reliable water sources considering quality and quantity,
   d. environmental problems that deserve prompt attention,
   e. motivation of the community toward a more rational use of the services,
   f. economic feasibility.
6. Water rates should be established in conformity to the IADB's current policy and criteria.
7. Every water and wastewater project should include a program to incorporate house connections and water meters for new users. Currently, external finance of the project covers the initial number of house connections and water meters. But, the utility should demonstrate to the Bank's satisfaction, its program for future users incorporation including, not only the connections and water meters, but the so-called basic house unit that in essence is a simple module (lavatory, water closet, shower).
8. In every water project, the source of supply should be reliable and its capacity should be determined based on evaluations performed over an adequate period of time, in order to satisfactorily determine variations in quantity. The expected reliable flow capacity of water wells should be established before the project's design. In addition to these criteria, surface waters should be treated to conform to the WHO's international water quality standards, and unaccounted for water losses in the system (that currently may exceed 50% of the production capacity; should be reduced to a 20 to 30% level).

9. Wastewater projects should include the design of a wastewater treatment plant (WWTP) to produce an effluent based on the water quality of the receiving waters. Reuse of wastewater is encouraged. Given the high cost of wastewater projects for small cities, low cost technologies should be chosen. Condominial and small bore wastewater systems should be considered and compared with traditional ones to obtain the least cost alternative. For rural areas *in situ* (latrine disposal to the soil) excreta disposal is considered adequate. Land required for the construction of water and wastewater works should be available before the construction stage begins. Ability to establishing rights-of-way should also be clearly assured by the borrower to facilitate maintenance.

Environmental Policy

IADB approved its environmental policy in 1979 establishing the criteria as well as the internal mechanisms to monitor the policy compliance. A Secretariat was created and proper instructions to the field offices were given. Other international lending organizations followed suit. The objectives of the IADB environmental policy are to:

1. Ensure the environmental quality of all the IADB's investment projects;
2. Incorporate the environmental component into project's design and help member countries comply with the policy including provision of economic resources for technical cooperation as needed;
3. Assist the member countries to identify environmental problems and concur in solutions. Organize international forums and promote ample discussion on the subject;
4. Reinforce local capabilities of the member countries encouraging the creation of government agencies (in some countries at Ministry level) with sufficient skilled personnel and resources, and fully devoted to deal with environmental problems;
5. Promote scientific investigations about the subject and allocate economic resources for that purpose.

Basic criteria and methodology for the environmental component of the investment projects are:

1. Early identification of the potential environmental problems and incorporation into the project's design.
2. Classification of the environmental impacts in four categories:
   a. Category 1. Projects where the environmental impacts are beneficial to the environment.
   b. Category 2. Projects where the environmental impacts are neutral to the environment.
   c. Category 3. Projects generating moderate environmental impacts and negative impacts for which there are known mitigating technologies and solutions such as wastewater treatment plants.
   d. Category 4. Projects generating negative environmental impacts, for which mitigation measures should be designed, implemented and monitored, such as dams, hazardous waste disposal sites.

Documents to be prepared by countries as part of the loan request to comply with the policy are:

1. Environmental Plan including the methodology for environmental monitoring
2. Environmental Impact Assessment (EIA), as a document of public domain for projects falling in categories 3 and 4.
3. Environmental Summary.

MAINTENANCE POLICY

This policy was approved in 1982 and has the following objectives, to:

1. Promote in the member countries the adoption of best maintenance practices;
2. Ensure the life of the built infrastructure and increase its efficiency;
3. Provide economic resources to help the member countries to comply with this policy. Technical cooperation and financing of rehabilitation and maintenance programs should be available;
4. Request from the borrowers the submission of an annual maintenance program for 10 years after completion of physical works.

Within this policy, maintenance is divided into two classes:

1. Periodic maintenance, that includes investments made from time to time to correct or prevent failures affecting the infrastructure. Currently, these investments might be part of a loan proposal.
2. Routine, including preventive, maintenance that includes investments made by the local utilities to ensure the efficiency and conservation of the infrastructure. These expenses are part of the water rate.

Water Rates

The Public Utilities Policy approved in 1996 supersedes and replaces the Public Utilities Tariff Policy in effect prior to 1982. The objective of the new policy is to promote the provision of public utility services that contribute to the long term economic development of the region and to the well being of its people by adopting a sector structure and regulatory policy seeking to:

1. Ensure the long term sustainability of the service;
2. Achieve economic efficiency in the provision of services;
3. Safeguard the quality dimensions of the services;
4. Promote the accessibility of the services to all citizens;
5. Meet wider national objectives, in particular the protection of the environment.

The basic conditions that must be met to assure the accomplishment of the objectives are:

1. The separation of the roles of policy formulator, regulator and entrepreneur;
2. The existence of a sector structure that fosters economic efficiency and maximizes the scope for competition;
3. The adoption of a sound and adequate regulatory regime;
4. The appropriateness of the institutional vehicles for regulation;
5. The adequacy of the legal framework;
6. The adoption of governance modes that provide efficiency incentives for management;
7. The existence of a firm government commitment to the objectives of the policy.

For the purposes of item number 3, above, two criteria are described in detail:

1. Prices for natural monopolies are regulated to assure long-term financial sustainability and to attain economic efficiency when competition is possible. Regulators should work toward the creation of conditions that lead to adoption of rates that signal the marginal cost of the service to the end user. In the case of public utilities, adoption of rates signaling marginal costs while maintaining the long run financial viability of the utility with an adequate return, must be a goal to be achieved.
2. Subsidies and/or other forms of intervention as a mechanism to achieve wider national objectives relating to social equity and environmental preservation may be considered in some cases. Well designed social policies should consist of the following elements:
   a. An explicit statement and clear justification of the chosen social objective is included.
   b. A procedure ensuring a mechanism for collecting the necessary funds, whether from general taxation or rate revenues is defined.
   c. A transparent mechanism for allocating funds to the target group is included.

As a summary, it is noted that IADB is a dynamic institution always responding to the needs of member countries. The policies stated above, were approved several years ago and some of them have been somewhat modified. But, in the end, the basic concepts of the policies remain after 45 years of successful IADB activity.

Chapter Three

Evaluation of Water and Wastewater Investment Projects

Evaluation of Water and Wastewater Investment Projects

THE APPRAISAL OF INVESTMENT PROJECTS IS A COMPLEX EXERCISE THAT REQUIRES THE PARTICIPATION OF A MULTIDISCIPLINARY TEAM WITH SUFFICIENT EXPERIENCE AND KNOWLEDGE. The team usually includes four specialists: a professional sanitary engineer (currently as Team's leader) (See Appendix I for a partial list of projects leaded by the author), an environmentalist, an institutional and financial expert, and a socioeconomist. The criteria and methodologies to appraise investment projects in Latin America are presented below. The evaluation areas of interest are technical, environmental, institutional, financial, and economic, each discussed separately in this Chapter.

TECHNICAL EVALUATION

Introduction and Evaluation Criteria

Introduction. The purpose of the technical evaluation is to verify if the project has been properly prepared based on international current practices, that sizing is adequate to attend the needs of the population to be served and finally, that all possible alternatives have been evaluated and compared in order to obtain the minimum cost alternative. The principal guidelines (checklist) for a technical evaluation are discussed based on the author's experience in Latin America.[2][9].

The levels of studies for an engineering project are, with each successive level representing increasing detail and improved cost estimates:

a. Preliminary level,
b. Prefeasibility level,
c. Feasibility level and
d. Final design level.

The preliminary level contains largely qualitative information used to present the nature and scope of the project and a gross estimate of the investment. It serves to reach a decision about whether to continue allocating resources, and to define a timeframe for the next, and subsequent stages of the project. The gathered information serves to prepare a project's profile although it is insufficient for a complete appraisal.

At the prefeasibility level, a series of alternatives are presented, analyzed and discussed taking into account not only the technical aspects but the socioeconomic and environmental aspects as well, to define the "least economic cost alternative" and its implementation stages.

For every alternative, a preliminary Environmental Impact Assessment and a basic design should be prepared. The principal components of the project are defined, such as source of water supply, intake location, scope of the network, capacity, treatment processes for water and wastewater, and definition of the final disposal point. In the selection of the least cost alternative the following criteria should be applied:

1. cost and benefits should be estimated based on frontier prices;
2. discount rate should be 12%;
3. the selected alternative must comply with the following requisites:
   a. to be the least economic cost, at present value,
   b. to have the highest net present value (NPV),
   c. to have the highest economic benefit in terms of NPV and economic internal rate of return.

The comparison of alternatives requires that projects at the prefeasibility level possess a cost estimate accuracy in the range of ±30%.

At the feasibility level, all the technical, environmental, economic, institutional and financial aspects should be included. All the necessary complementary technical studies should be completed. The cost estimate should be in the range of ±20%.

Once the feasibility level analysis is completed the Investment Master Plan that represents the "Minimum Cost Expansion Plan" can be prepared. The Master Plan is necessary to implement the selected alternative by stages with a time horizon of 20 years.

At the final level, the project should include:

1. description of the project,

2. detailed plans including structures, electricity, architecture, instrumentation, mechanics, ancillary works, plumbing etc.,
3. general and particular specifications for materials and equipment,
4. construction quantities, unit prices and cost estimate in the range of ±10%,
5. tender/bid documents.

Some good references to help in project's preparation can be found in the following handbooks jointly prepared by PAHO and IADB[(8)]:

- *Water and Wastewater Cost Analysis.* Dr. Edward Martin
- *Review and Formulation of Design Criteria.* Dr. Donald Lauria
- *Appropriate Technology.* Dr. Edward Martin

Technical Criteria for Water and Wastewater Alternatives Studies. The feasibility of implementing a water and/or wastewater project can be determined after a complete alternatives study is done. The main technical criteria used to conduct an alternatives study are:

1. The project should, at pre-feasibility level, have the principal components clearly defined and sized, and possess a proper cost estimate;
2. The same design period and stages should be used for all alternatives in order to normalize the alternative comparison;
3. The source of water supply should be defined after an economic comparison is made between surface and groundwater sources;
4. Transmission lines should be designed with sufficient capacity for the end of design period, e.g., twenty years, but, an economic comparison should be made to determine whether a transmission line built in 2 stages of 10 years, is more economical;
5. Alternative water treatment plants should produce the same water quality for a satisfactory comparison of costs. There must be a comparison made among several technologies;
6. The water network should be designed using digital models. There is no limitation on the number of schemes compared while seeking the least cost alternative. Currently, the network component that represents about 70% of the total cost of the project, must be carefully analyzed;

7. Alternative wastewater treatment plants should produce the same effluent quality to arrive at true cost comparisons. There must be a comparison among several technological alternatives as in the case of water supply. Sludge treatment and environmentally sound disposal should be included. The wastewater network should also be designed using digital models. The drainage basins should be clearly defined. It is important to gather geological information along the main collectors, interceptors, pumping stations and treatment plants. For a recent project in Panama, 18 alternatives were compared, considering different wastewater treatment technologies;
8. A catalogue of unit prices should be prepared based on current local prices as well as international prices for external components of the projects. Some countries already have this catalogue, but it may have to be updated to current conditions. The technical review should be made coordinating the efforts of engineering and economic evaluations.

Appropriate Technology. D. Okun in his book[17] referring to "Inappropriate Technology," says: *"The aftermath of implementation with inappropriate technology is especially noticeable in the treatment units of water plants: coagulation and rapid mixing, flocculation, sedimentation, filtration and disinfection."* Several examples from the developing world are cited. He concludes his discussion about saying: *"Rather than transferring technology from the industrialized world to the developing world, engineers from the industrialized world might well learn something of simplified practices that have been found satisfactory in the developing world, so that, as they provide assistance to other countries in the developing world, they would be using technologies that are appropriate and that can be easily operated and maintained."*

E. Martin in his book[18] says about appropriate technology, and based on his Latin American experience: *"The word 'Appropriate' may mean many things, especially in the context of technology applications, as in this case for development and management of water and wastewater-quantity and quality. The technology is 'Appropriate' if it is suitable to the functional application for which it is intended from the viewpoints of cost, operability, and simplicity of design."*

The author, in his presentation in Cuernavaca during a Symposium about "Cost Efficient Technologies in Water Treatment,"[16] said: *"The main concept in the allocation of economic resources is the cost minimization that implies the alternatives exercise to select the least cost alternative."* It is recognized that the cost of a treatment plant is relatively minor in comparison with the total cost of a water system where the distribution network may represent about 70%.

Water treatment is very important from a health standpoint. In regard to water treatment there are 3 alternatives: conventional design, patented plants, and plants incorporating new technological advances that have been developed in Latin America. CEPIS/PAHO, has made an excellent contribution to the development of new treatment technologies since the decade of the 70's and deserves credit for the achievement, that may be called "Latin American Appropriate Technology." Several demonstration projects were carried out and several publications on the subject have been distributed all over the world[12][13][14][15]. The IADB has been carefully monitoring these developments because the purpose of the investigation coincides with the IADB's policy for allocation of economic resources. Furthermore, use of appropriate technology reduces cost, simplifies operation and maintenance, utilizes local capabilities, labor and materials, and reduces the dependence for the acquisition of machinery from the industrialized world. All of these features impact the cost of water, making it more affordable to the consumers.

Water Supply

The source of water can be from surface or groundwater, whichever is most suitable and economical, taking into account the following factors: adequacy of yield, suitability for use—that is water quality and degree of treatment, conflicting use(s) due to water rights that belong to others and competing water uses.

Surface Waters. Rivers, lakes or impounding reservoirs can be used as surface sources. The source of supply should be selected taking into account the impact of potential sources of pollution on the watershed, like: size of population, human and animal wastes, agriculture or industrial activities. Characterization of the raw water quality should be done; extended physical, chemical and bacteriologic analyses should be performed.

The concentration time of runoff is dependent on the slope of the ground surface, surface cover or vegetation, and dimensions of the watershed area. The determination of the stream flow would be based on stream flow records (if stream flow records are unavailable, the records of nearby streams having similar watershed characteristics may be used), and rainfall records supplemented by adequate hydrologic calculations. The minimum stream flow should be estimated when water is withdrawn and consideration given when the stream is used for wastewater disposal. The maximum stream flow should be considered when designing reservoir spillways, stream channels and flood control works, drainage systems, pumping stations, and when providing safe elevations for facilities near streams.

The surface water source development should use the base flow (minimum sustained dry weather flow), developed using a hydrograph(s), and including periods of little or no precipitation. The potential yield of the stream, neglecting losses due to evaporation and seepage, can be estimated as shown in the following table:

Coefficient of Variation of Mean Annual Flow	Maximum Practical Yield as % of Mean Annual Volume
0.25	95
0.50	80
1.0 +	30-80

In summary, in order to select an appropriate source, data are necessary for hydrologic, geologic, topographic, water quality, and type of watershed uses. To estimate the safe minimum yield, extensive measurements should be carried out in rainy and dry seasons. To determine the components of a project using surface water three conditions should be analyzed.

1. the base flow of the source is always greater than the water demand for the maximum day. No impounding reservoir is needed
2. the base flow of the source exceeds the water demand for the maximum day part of the year. In such a case, an impounding reservoir is needed.
3. the base flow of the source is always less than the water demand for the maximum day. In such a case a larger impounding reservoir is needed, than for "2." above.

Impounding Reservoirs. Storage is required when the base flow of the stream is less than the required rate of withdrawal as indicated above. Storage capacity can be calculated from a "mass diagram" and should consider the following requirements: maximum rates of seepage and evaporation, drought periods, reservoir silting and demand for multiple uses of stored water. The location of the impounding reservoir(s) depends on the storage capacity required, topography and other factors, including environmental factors. In the world, there are many examples of the negative environmental impacts generated by impoundments. Special care should be taken in design of impoundments to eliminate or mitigate adverse impacts.[11]

Groundwater. Groundwater development should be based on aquifer capability, geologic character of the aquifer and the strata above and below, aquifer discharge, aquifer recharge and aquifer properties such as permeability, transmissibility, porosity, specific yield and water quality.

1. *Springs.* Springs are very common sources of water supply in several areas in Latin America. Springs may be characterized and classified as thermal, gravitational, depressive, artesian, tubular and fracture. Thermal springs are generally not suitable for water supply due to high mineralization/solubilization. Suitable collection works for springs are as follows: collection chamber duly covered for gravitational springs, open trenches for depression, buried pipes for depression, and wells for artesian springs. Usually, disinfection treatment is applied to guarantee microbiological water quality.
2. *Infiltration Galleries.* Infiltration galleries are common sources of water supply in several areas of Latin America. The most important water supply source for Lima, Peru, has been the infiltration galleries located upstream of the surface water treatment plant called, "La Atarjea." Infiltration galleries are usually constructed utilizing buried perforated pipe packed in gravel. Design velocities should be about 0.60 m/sec, or less. The water should be collected in a covered sump, to protect water quality. Usually, disinfection treatment is applied to guarantee microbiological water quality.
3. *Wells.* The ability to yield water can be determined by test wells. The IADB's policy requires that wells are drilled and tested for at least 72 hours before estimated capacity is considered safe to be used for design. Wells may be of several types and constructed by several methods such as: shallow dug well, bored wells, driven wells, jetted wells and drilled wells. Again, disinfection treatment is usually applied.

Design Period. The design period is dependent on the expected life of the structures, potential for expansion of the system, type of demographic growth considering industry, and cost of money. Commonly, a 20-year design period is adopted with 2 stages of 10 years each.

Population Forecast and Densities. Based on the available population statistics and using different methodologies to compare results, the trend for future population and area densities by project stage is determined. Several methods are used to make a proper forecast of the population, such as: rational, graphical, arithmetic, geometric (Malthus), variable increments, comparative, logistic (Vershul) [more applicable to old populations with low growth rate like Montevideo, Uruguay], and logarithmic parabola (Reed), among other techniques.

Water Consumption. The factors that affect water consumption are, among others, weather, standard of living, existence of wastewater collection system, type of population, water cost, existence of private water sources, water quality, pressure, extent of industrialization, coverage with water meters, and management of the system. Based on current statistics from water utilities and common practice in the region, the per capita water consumption is established. This value should not be confused with the term "demand," that is water consumption at different prices. Currently, common figures adopted for design are:

1. For urban water projects, 250 to 300 liters per capita per day (including an estimate of 20 to 30% of unaccounted for water. Industrial consumption is considered additional.
2. For rural water projects, 60 to 120 liters per capita per day are common numbers. The reduction in the parameter of consumption is due to the different level of service adopted.

Currently, rural water projects for small towns do not have in-house connections and meters. The size of the water distribution line, commonly 1 to 2 inches and one connection per house, are probably not accommodating of in-house extensions. The owner is responsible for fetching water from one source within the house. In certain cases "easy access" to water consists of a set of standpipes centrally located in the community but external to homes. People must walk to the standpipes and carry their water home, and a result, consumption is low in these cases. For large rural towns the distribution lines are 3 to 4 inches in size, which in turn allows multiple within-house connections and eventually, water meters. Consumption values are higher. Currently in Latin America the population limit to consider a town 'rural' is 2,000 people. Nonetheless, in some countries like Argentina the limit is from 6,000 to 10,000 people.

Variation of Consumption. The consumption for the maximum day flow rate is often adopted at 120% of the annual average consumption. For hourly variation or peak flow rate consumption 200% of annual average consumption may be used. The peak consumption factor is related to the population in an inverse relationship shown below. According to Harmon, this relationship can be expressed by the following formula:

$$P_f = 1 + \frac{14}{4 + P^{1/2}}$$

Where:

P = Population (thousands)

P_f = Peak factor

For example, the peak factor for a town with a population of 2000, is given by:

$$P_f = 1 + \frac{14}{4 + \sqrt{2}} = 1 + \frac{14}{4 + 1.41} = 3.59$$

As the population increases, the peak factor decreases.

There is another formula used in Latin America's practice[9] to estimate the peak consumption:

$$Q_p = 1.5 + 2.5 / Q_{md}^{\ 1/2}$$

Where:

Q_p = Peak consumption (liters/sec).

Q_{md} = maximum day consumption (liters/sec).

Unaccounted For Water (UFW). In urban water projects in Latin America the unaccounted for water represents about 50% of the water produced. For design of a new project a provision of 20% to 30% unaccounted for water should be included as a reasonable figure to be achieved in well-managed water systems. In developed countries this figure is around 10 to 15%. The term UFW (Non Revenue Water) includes the physical losses in the system as well as the commercial losses due to errors for instance in water meters readings. Commonly, the physical losses account for about 70% of the total losses. To evaluate unaccounted for water in an existing network it is necessary to isolate sectors with 100% house meters coverage. Then, using master flow meters the water input to the sector is measured as well as the water consumed by the households. The difference is unaccounted for water. In Latin America most of the networks are partially metered making it more difficult to estimate the UFW (see Appendix II).

Regulation Storage. In water projects a volume of 20 to 25% of average annual consumption is included as storage to provide for hourly variations in flow. This storage can be laid in the below ground or elevated structures.

Fire Protection. In urban water projects in Latin America, a practical criterion for fire protection is used, based on 15 liters per second per fire hose, and two hoses working simultaneously for two hours. For specific cases to protect valuable properties, the National Board of Fire Underwriters (USA) formula can be applied:

$$Q = 1020 \times P^{1/2} \times (1 - 0.01P^{1/2})$$

Where:

Q = Flow Rate, gallons/minute
P = Population in thousands
The fire protection flow rate increases with increasing population.

Pressure. In water projects in Latin America, a minimum pressure of 17.5 meters with a maximum pressure of 50 meters is currently used for design (25 to 72 psi).

Distribution Network. The capacity of a distribution network for a water project is estimated based on two conditions, whichever gives the higher value: the maximum hourly consumption (peak), or the maximum day plus fire protection. For urban water projects, the network design uses the Hazen-Williams formula and the Hardy Cross grid theory commonly using a digital computer model. For rural water projects a tree network is considered in the hydraulic analysis. Common materials for pipes in Latin America are: ductile iron, PVC, fiber-cement and asbestos-cement. IADB recommends against use of asbestos-cement pipe due to the potential health risk from leached asbestos in certain waters.

Pumping Stations. For planning and design of a pump station the following considerations should be taken into account:

1. Purpose of service. The design of the pumping station depends on its use for transmission from the water source, pumping station within

the distribution system, pumping to elevated storage, pumping for fire protection, booster pumping and pumping at treatment plants;

2. Piping layout;
3. Demand requirements;
4. Static lift requirements: suction head, discharge head;
5. Power available and reliability;
6. For pumping capacity multiple pump installations should normally be used to meet the maximum demand where there is no elevated storage, and maximum day demand where there is elevated storage;
7. Selection of pumps should be based on service requirements, cost, service life, maximum and minimum head conditions. Catalog information from manufacturers can be of great assistance. The most common type of pumps are: centrifugal, and turbine for wells.
8. Installation requirements should consider the following aspects:
   a. location;
   b. except for submersible pumps, the pumping station should not be subject to flooding;
   c. provision for a positive suction head;
   d. piping arrangement to provide minimum interruption in suction piping, static lift should be less than 4.5 meters and in all conditions, the suction lift must have positive priming;
   e. valves (gate, check and relief) at pump stations should be arranged to permit pump removal without service interruptions;
   f. flexible couplings should be provided to relieve strain and to take up misalignments.
9. The type of power selected should be based on several factors including dependability, availability and economy.

Water projects using wells are very common in Latin America. Pumping stations are designed for the maximum day unless they pump directly to the network, and often work about 18 hours per day. An underground reservoir is needed for storage. Usually the pumping stations have units in stand by. In some places they also have stand by energy units when the public electric service is not reliable.

Transmission Lines. Size of water transmission lines is determined by comparing cost, different pipe sizes, and operational costs (mostly energy).

The 'economic size,' is the size where the cost obtained, combining the cost of pipes and the costs of operation, is a minimum. Common materials for pipes are: ductile iron, steel, fiber-cement and concrete. A practical formula to estimate the size is:

$$D = 1.5 \times 2.54\, Q^{1/2}$$

Where:

D = diameter in centimeters

Q = Flow, liters/sec

Water Quality. Quality of the raw water source should be determined by examination of the bacteriological, physical, and chemical characteristics. Highly contaminated waters should be rejected in the initial analysis of a project because of the high cost and difficulties of treatment. Extended sampling and analysis should be performed to characterize water quality. Water quality for drinking purposes must meet the guidelines established by the World Health Organization. These limits are incorporated into the local standards in most of the countries in Latin America. After treatment, the water should be free of disease-producing organisms. The E. coli Index is the common indicator of contamination by pathogenic organisms. In finished water the standard indicates that the E. coli MPN should be near zero and always lower than 2.2. Physical measurements give an indirect indication of undesirable substances and gauge the palatability and acceptability to consumers. Chemical analysis gives a direct measure of pollution by substances that are toxic, non-palatable or otherwise objectionable. Microbiological determinations yield information on pollution by sewage and potential tastes and odors (by algae).

Capacity of the System. The size and scope of water and wastewater projects are obtained by combining the future population estimates over the life of the project and the per capita water consumption figures.

Water Treatment Plants. Water treatment plants are almost always required in order to comply with the drinking water quality requirements. Turbidity, color and bacteria are removed by coagulation, sedimentation filtration and disinfection, Tastes and odors are removed by aeration, carbon and chlorination. Hardness is removed by ion exchange processes. Iron and manganese are removed by aeration, filtration and softening. In Latin America some sources, like Argentina and Chile raw water sources, require respectively, the reduction of fluorine (to safe levels required for dental hygiene), and arsenic levels by special treatments. Commonly, a water treatment plant consists of the following treatment processes:

1. pretreatment, to reduce high concentrations of solids as in the Lima, Peru treatment plant, where the turbidity exceeds 20,000 JTU (or NTU), during the rainy season. Roughing filtration is used as pretreatment. Sometimes, roughing filtration is adequate and is the most appropriate treatment for small communities,
2. rapid mix; flocculation and coagulation,
3. sedimentation,
4. filtration,
5. other processes such as aeration, heavy metals reduction, odor control, and,
6. disinfection, currently by chlorine (Ozonation is not a practicable treatment in Latin America due to its high cost compared to chlorine, except particular applications found in Brazil). When the source of water supply is groundwater, the only required treatment may be disinfection. Further discussion on treatment will be concentrated on the most current processes and the most appropriate technology.

Brief Review of the Unit Processes for Water Treatment Plants. Most of the information presented here comes from references (17), (18), (19), (21), (22), and (23). The use of pilot plants in the design process is always recommended unless data are available from previous studies. The required dose to apply a coagulant for water treatment (mix and flocculation) is impossible to calculate using analytic procedures. The "jar test" has been developed to measure, in the laboratory, the proper dose as an approximation to the situation faced in the full-scale plant. Monitoring of filtered water measures effectiveness of removing turbidity. Pilot plant studies take into account the variations in quality of the raw water and the expected quality after treatment. Time devoted to these studies and economic resources allocated are worthwhile.

Rapid/Flash Mix[(10)]. Rapid mix, or flash mix, is used to generate a mixture of water, contaminants and coagulants, and to promote coagulant dispersion. This dispersion should be homogenous, and equally and quickly achieved. Several devices can be used for rapid mixing: pumps, Venturi flumes, Parshall flumes, Palmer-Bowlus flumes, weirs, hydraulic jumps, air jets, paddles, hydraulic baffles, turbines or propellers. It has been recognized that the coagulation is the heart of water treatment. It is important to apply the coagulant in the proper dose and in the most efficient manner. A common alum dose is 25 to 35 ppm. Proper mix and flocculation are the key elements to impact positively on the sedimentation and filtration unit processes. The principal parameters for design are:

1. detention time: 1 - 2 minutes;
2. mixing intensity (velocity gradient): G = 700 to 1000 sec^{-1}
3. power input = 26.3 to 52.6 $kw/m^3/sec$

The velocity gradient (G) is a measure of the mixing intensity in mixing and flocculation processes. But the G values for flocculation (in the order of 50 sec-1) are much lower than for rapid mix. The G value may be referred to as the root mean square of all velocity gradients in a given chamber. We cannot measure velocity gradients conveniently but we can measure the energy input to a device.

Flocculation. Mechanical stirring or slow paddle mixers gently stir a mixture of water and coagulant to form the floc. During the agitation the particles collide increasing the size of the floc particles; larger particles are easier to remove by settling. Larger floc is more likely to break up however, so a series of chambers is used to reduce the initial velocity while approaching the sedimentation unit. Flocculation units can be either of horizontal or vertical flow. In Latin America old plants use horizontal flocculation basins and hydraulic baffles. Hydraulic flocculators once in place, are rigid and it is difficult to modify velocities when needed unless, the baffles are removeable. Consequently, in many plants in Latin America, vertical turbine pumps are preferred (see table below for large plants). The principal parameters for design are:

G: 10 to 100 sec^{-1} · Flocculator detention time (t) is usually 10 to 30 minutes; accepted range of values for Gt: 10^4 to 10^5

A partial list of plants operating in Latin America using hydraulic or mechanical flocculation are shown in the following table.

Name and Location	Capacity m^3/sec	Hydraulic Flocculation	Mechanical Flocculation
Viscachitas, Santiago, Chile	10.0	x	
Madrigal, Santo Domingo, Dominican Republic	12.0	x	

Name and Location	Capacity m³/sec	Hydraulic Flocculation	Mechanical Flocculation
Cutzamala, Mexico City, Mexico	10.0		x
São Paulo, Brazil	5.0		x
Aburra, Colombia	6.0		x
Region Central, Venezuela	7.5		x

The author's first professional experience in plant operation was many years ago in Peru, when he had the task to put into operation a small newly built plant. The plant included the following unit processes: rapid mix (Parshall flume) with alum liquid feeder, baffled channel horizontal flocculator, sedimentation, filtration and chlorine. The optimum chemical dose was obtained from jar tests (after a series of numerous trials). Once the alum was applied to the mixing chamber, the floc started to form immediately and increased in size throughout the flocculator and settled before reaching the sedimentation unit. Also, at the entrance to the sedimentation unit, the larger floc particles strongly collided. The solution was to make physical modifications to the plant, including: increasing the width of the turn-around at every baffle, to make its width 1.5 times the width of the previous channel, and increase the width of the last channel connecting both units. In current practice in many plants, it is desirable to locate the flocculator and the sedimentation unit in one single chamber to make the transition between them smoother.

Camp has provided the basic equations for calculating the rapid mix and flocculator chambers. A simplified presentation follows[21]:

$$\text{Rapid mix, } G = 425\left(\frac{Hp_w}{t}\right)^{1/2}$$

Where:

G = energy input, sec^{-1}

Hp_w = water horsepower per mgd

t = residence time in minutes

$$\text{Baffled chambers, } G = 178\left(\frac{H}{t}\right)^{1/2}$$

Where:

G = energy input, sec^{-1}

H = head loss in feet of water

t = residence time in minutes

Assumed: Viscosity of water at 4o°C. For other temperatures a correction factor should be applied.These equations are used to design flocculation units. In reference (17) there is a presentation about the theory to design a flocculator including a useful numerical example.

Sedimentation. Sedimentation removes the particulate matter following classic laws of Newton and Stokes. Once the floc particles are big enough they precipitate by gravity to the floor of the sedimentation basin. The basins can be circular, square or rectangular. Due to the fragility of the floc following formation, the design of the inlet structure is important. Special care should be placed on the design of both the inlet and outlet arrangements. In rectangular basins the major portion of the particulate matter is concentrated in the first third of the length of the sedimentation basin. The sedimentation basins can be cleaned mechanically or by hand. Sludge removal is mechanical collection followed by pumping from the basin center (in the case of circular basins) or, from a depressed end (in the case of rectangular basins). The sedimentation is a vitally important step in treatment of surface waters and in most cases is the determining factor for the overall cost of treatment. Poorly designed sedimentation units will result in reduced treatment efficiency. The major classifications of sedimentation units are: horizontal flow, upflow, inclined plates and tube settlers. For horizontal flow the common design parameters are:

1. detention time: from 2 to 6 hours
2. surface loading: from 6 m^3/day/m^2 for plain sedimentation without chemical coagulants, to 60 m^3/day/m^2, for alum coagulation.

Hazen in 1904 and Camp in 1946, demonstrated that the efficiency of sedimentation is independent of the depth of the tank. The surface loading, m^3/day/per m^2 of surface area of the tank is the main design parameter. Based on this finding, the use of tube or plate settlers is a "new" development that results in improved efficiency of sedimentation basins. Tube and plate settler design is based on the fact that settling efficiency is independent of depth. Tube and plate settlers usually include proprietary designs and details are available from manufacturers. However, in several locations in Latin America, informed operators have converted existing rectangular sedimentation tanks by adding plywood sheets to the tanks in a sloped configuration between the inlet and outlet of the tank. The result is to provide a significant increase in treatment capacity without the addition of new settling tank volumetric capacity (increased settling efficiency in existing tanks). This technology may be successfully used to increase the capacity of old plants. In Buenos Aires, at the Palermo water treatment plant, the inclined plate technology was applied to old existing sedimentation tanks with an increase in capacity of 10 m^3/sec, with a minimal investment of US$15 million. Coagulation aids (like the application of polymers) may increase the efficiency of the sedimentation units also.

Azevedo and Ritcher[24] from Brazilian experience, present the following table for sedimentation parameters, using what they call "innovative technology"[12].

Characteristics of the plant	Surface loading (m^3/day/m^2	Horizontal velocity Max. (cm/sec)	Detention time (hours)
Small plants with limited operator experience. Conventional design	20- 30	0.4 - 0.6	3-4
Plants using recent innovative technology and good operator experience.	35- 45	0.7 - 0.9	2- 3
Large size plants using coagulation aids and very good operator experience.	40- 60	0.6 - 1.25	1.5- 2.5

Upflow solids contact sedimentation units (UASB), combine in a single tank, mixing, coagulation, flocculation, solids separation and sludge removal. The inlet is located at the bottom of the tank and its velocity is gradually reduced upward due to the increase of the cross-sectional area of the tank. Influent turbidity should not be more than 50 JTU. In Brazil, there are many plants of this type. The addition of tube settlers or plate settlers increases the allowable surface loading on these units. Surface load may be in the range of 36 to 100 m^3/day/m^2.

Filtration. The main aim of filtration is to improve removal of turbidity, color and bacteria, beyond what is possible following sedimentation. The mechanisms operative in filtration units are physical, chemical and in sometimes biological (nitrogen removal). There are two conventional kinds of filters depending on flow rate capability and other features.

The slow sand filter consists of a layer of ungraded fine sand. The water passes through it at reduced loading rate, from 2.4 to 7.7 $m^3/day/m^2$. An important component of the slow sand filter is the creation of a thin layer at the surface of the bed called "schmutzdecke." This layer is composed of a large variety of biologically active microorganisms that assist in the filtration process.

The slow sand filter has significant potential for application in small cities in rural areas; service for large cities would often be prohibited due to low flow rate and large area of land required. A limitation is that it works efficiently with turbidity less than 50 JTU. Cleaning is done by scraping the surface of the sand, which is externally washed and returned to the filter. A unique slow sand filter called "dynamic filtration," originally developed in Russia, has been used successfully in several rural programs in Latin America such as Honduras and Nicaragua C.A.

The rapid sand filter has a layer of graded sand (effective size 0.25 to 0.50 mm and uniformity coefficient 1.25 to 1.5), located over a layer of gravel. In some cases, an anthracite layer is put on top of the filter to increase the flow rate. This kind of filter is called a dual media filter. Use of dual media can increase the flow rate by 200%.

According to J. Azevedo and C. Ritcher the first rapid filter in the world was installed in 1880 in Campos City in Brazil. Rapid sand filters can be classified according to their characteristics such as: type of filter media, rate control feature, direction of flow, the classic down flow and the upflow or both, form of operation, gravity or pressure driven. The basic components of a rapid filter are: filter media, scour wash feature on top, special underdrain filter bottom, backwash and rate control. In the past 20 or more years, the technologic development in numerous investigations was directed toward increasing the flow rate from the conventional 120 $m^3/day/m^2$ to higher rates, reduction of the size of the plant, eliminating rate control, reducing cost and simplifying operation. The flow rate can be increased easily to 300 $m^3/day/m^2$ or more, depending of the result of pilot studies.

For waters with low turbidity, of the order of 20 JTU, and minor variations all the year round, direct filtration and disinfection, is the appropriate treatment. Filtration and disinfection are applied, for instance, to the water from Gatun Lake, one important source of water supply for Panama City, Panama.

Disinfection. Disinfection is the final step in the treatment train that guarantees microbiological water quality. Commonly, chlorine is used for this purpose; chlorine also helps in the reduction of hydrogen sulfides, iron, odors and flavor. Chlorine is currently used in gas form and also as sodium hypochlorite for small applications. Due to the possibility of the formation of chlorine disinfection by products (DBPs); ozone is used in the industrialized world but it is not affordable for most of the Latin American countries. Chlorine dioxide, does not have the same ability to form DBPs, is used in some applications though is more costly because it has to be produced at the water treatment plant.

Cost of Water Treatment Plants. There are many references for cost estimating. Costs should always be modified to the current year. In 1981, J. Perez published a report related to the Research Program carried out by CEPIS, to develop "Appropriate Technology" for water treatment in Latin America. The research included the following tasks: diagnosis of the treatment strategy in more than a hundred plants in Latin America, with capacities from 6 lts/sec to 60 m^3/sec, dissemination of the findings through the publication of books, application of developed technological criteria to the design of new small capacity plants like those located in Cuenca, Ecuador, El Imperial in Peru, Cochabamba in Bolivia, Prodentopolis and Linhares in Brazil, and several rural water projects in Argentina, Ecuador and Venezuela. As a consequence of these activities, CEPIS arrived to the following conclusions:

1. Existing conventional plants can increase their capacities with a cost 30 to 40% of the cost using conventional technologies.
2. The new water treatment projects using the technological advances have a cost 40 to 60% of the cost of conventional plants and 45 to 50% of the cost of patented treatment plants.
3. The O&M costs of the water treatment plants using technological advances are 10 to 15% less than the cost of conventional treatment plants.
4. The use of a new technology implies the processes optimization using high surface rates and as a consequence the water treatment plants are smaller and more economical.

5. The use of the new technology can be easily handled by the current technical capability of Latin America's countries.
6. There is no limitation for using the new technological concepts, neither for the size of the water treatment plant nor for the water quality.

Summary of CEPIS Research Program

1. The objectives of providing a water supply system include improving health, economy and development without adversely modifying the environment. An adequate and appropriate technological solution to attain the objectives should consider the use of local resources, and be relatively easy to construct and operate.
2. In order to solve the problem of water supply in Latin America, the following are technical alternatives: import from developed countries as has been done in the past, develop and apply simplified appropriate technology.
3. The research program conducted by CEPIS demonstrated that it is possible to design, build, operate and maintain water treatment plants using appropriate technology, resulting in high technical efficiency at a lower cost than other technological solutions.

Even though the research program recommends the use of high surface loading rates for the design of new water treatment plants, in the opinion of the author more conservative or 'average' design criteria values should be used in the design of new water treatment plants rather than extreme values. This recommendation comes from personal experience working for many years in Latin America, where, due to economic constraint expansion plans are almost always delayed.

Wastewater

Most of the information presented in this section comes from references (9), (27), (28), and (29).

Wastewater Flow. Only municipal wastewater consisting of domestic and industrial discharges is considered herein. Stormwater and combined sewer discharges are not explicitly addressed but may be prominent in many proposed projects. Wastewater flow may be estimated as a percentage of water consumption, e.g., 80% of water consumption. Industrial wastewater flow is included if the industrial water supplies are a part of the water supply system, otherwise industrial wastes must be considered as additional flows.

Wastewater Collection. The purpose of the wastewater network is to collect all the wastewater produced by households, as well as other liquid wastewater sources, and convey wastes to be properly treated before disposal. Once the wastewater is collected through laterals, collectors, trunks and interceptors it should be conveyed to one or more points for final disposal. The wastewater collection system is designed for peak flow (160% of water consumption), to which must be added infiltration/inflow rates, and other rates relevant in specific cases. Infiltration allowances adopted in Panama's Wastewater Feasibility Study were: for existing network, 0.15 liter/sec/hectare and for new pipes, 0.06 liter/sec/hectare. Another conservative figure recommended in several references is 700 to 2000 gallons per acre per day. Infiltration figures should be estimated based on field tests. Before the raw wastewater is released to the environment, it should be treated. In conventional systems, lateral lines are 6 to 8 inches diameter. Minimum slope for 8" sewers is 0.004 except in the origin of laterals were it should be steeper. It is usual to include some provisions for flushing water in this initial portion of the laterals. Velocity in the sewers is calculated using the Manning formula (in metric units):

$$V = \frac{1}{n} R^{2/3} S^{1/2}$$

Where:
V = mean velocity, m/sec
n = Manning's coefficient of roughness. Currently 0.0013 for concrete pipes
S = hydraulic gradient, m/m
R = hydraulic radius, m, the hydraulic radius is defined as the water cross sectional area, A, divided by the wetted perimeter, P. For a circular pipe, if D is the diameter of the pipe, then R is:

$$R = \frac{A}{P} = \frac{\pi D^2 / 4}{\pi D} = \frac{D}{4}$$

Minimum velocity should be 0.6 m/sec flowing full and maximum no more than 3.0 m/sec, to avoid excessive wear. To facilitate maintenance manholes should be provided every 50 m for smaller collectors and 100 to 150 m for larger sizes. Collectors are 10 inches diameter and up. The minimum velocity in the pipe flowing full, is the same 0.60 m/sec. Trunk sewers receive the discharge of the collectors and have much larger diameters. Sometimes due to its large diameter they can work with less velocity but not less than 0.30 m/sec. The last portion of a network is the interceptor sewer that conveys all the wastewater to the final disposal point/s.

Materials currently used for wastewater networks are: plain concrete, reinforced concrete, fiber cement, ductile iron. The election of the material is based on the strength of the wastewater and the quality of the soils.

Simplified Wastewater Systems. Condominial and small bore wastewater systems should be considered as alternatives to serve the low-income users of the cities. Here is a brief description of the principal features of each one:

Condominial systems[25], have been used in some countries in Latin America but they were introduced and redeveloped in 1981 in the North East of Brazil. The condominial system consists of 3 components: the public collection sewer, the treatment unit and the collective connection branch installed at the location of the house. Each house in the block connects to the collective branch through a small inspection box. The public collection system has sewers of 6 inches, the condominial branch is of 4 inches with few connections to the public collection system. The wastewater treatment plant is a simple one: stabilization pond or anaerobic filter. It is estimated that the cost of this kind of simplified system is about 50% of the cost of the conventional system. Principal advantages are: easy construction, less extension of the public collection system, low construction cost of the entire system, lower cost of operation. The disadvantages are: inappropriate use of connections due to discharges of solid materials and garbage. The success depends entirely on the attitude of the users based on good communication, explanation and training.

Small bore systems[25], can be considered as a combination of individual disposal system (on site septic tanks) and an effluent collection system made of small diameter pipes that receive and transport clarified wastewater free of suspended solids under pressure; a pressure does not require defined slopes. This low cost technology was developed early in the 70's in the USA and introduced to Brazil in 1979. The advantages are similar to the condominial systems, in effect they are simple, more understood by the users and the pipes can operate under pressure because they carry only water. The saving in cost is similar to the condominial systems. The only disadvantage is the clogging of the pipe due to the lack of cleaning of the septic tanks. Even though the systems are less costly public acceptance is not assured, because they consider they are receiving an inferior service – not the same service as other communities. In Uruguay, it was not considered as an acceptable solution for cities, even small cities.

Pumping Stations[10]. In wastewater projects the pumping stations, known as lift stations are designed for peak flow. The pumping station is usually designed in a wet well - dry well configuration unless submersible pumps are used. Pumps work in cycles of 10 to 30 minutes of retention time to avoid septic conditions. Centrifugal pumps are typically used. Positive-displacement screw pumps (Archimedes) are used when only an onsite lift (low lift) is required. Gate valves, check valves and devices to control water hammer should be provided on the pumping stations. The pumping stations are usually designed with stand by units. To make reliable its operation, proper instrumentation is required as well as standby energy units.

Treatment Plants. Wastewater treatment plants are necessary to reduce environmental pollution in the receiving waters (including marine waters) and to conform established water quality standards. The pollution elements of importance in wastewater treatment are: suspended solids, biodegradable organics, nutrients, refractory organics, heavy metals, dissolved inorganic solids and biological pathogens. Peak flow is used for hydraulic design to establish maximum water levels and freeboards. The average annual flow is used to design treatment processes. Minimum flow is used to determine the minimum levels for falls in weirs and channels and submergences on equipment.

Typical Wastewater Composition. A typical wastewater for medium strength composition in Latin America is presented in the following table:

Parameter and Units	Quantity
BOD_5 ppm	250
COD ppm	690
Total suspended solids (SS) ppm	290
Settling solids ml/L	9
Grease ppm	36
Nitrogen (as NH_3)	27
Nitrogen (organic)	18
Phosphorous (PO_4) ppm	20

Parameter and Units	Quantity
Total Coliform MPN/100 ml	10^7 to 10^8

Wastewater Treatment Processes. The common levels of wastewater treatment are: preliminary, primary, secondary and tertiary. But these frequently used designations are "terms of art," used to simplify discussions among professionals and are not expository about the specific level of treatment. A wide range of effluent quality is possible within each designation and thus the terms are not useful for predicting impacts on receiving waters. Modeling and receiving water quality are discussed later in this chapter.

Preliminary treatment eliminates floating solids, grit, grease and scum and generally includes screening, and grit and scum removal.

Screening. For preliminary treatment or pretreatment, racks, screens, comminutors, grinders and sometimes microscreens can be used. Bar screens with large openings 1.5 to 6 inches are used to remove very large trash before entering treatment processes. Some modern WWTPs use both racks and fine screens. Coarse racks of ¼ inch openings remove large solids, rags and debris. Fine screens of 0.05 to ¼ inches remove material that may increase operation and maintenance in downstream treatment processes. These units can be manually or mechanically cleaned depending on the size of the plant. Trash removed should be disposed in a landfill. Comminutors and grinders are used to shred suspended material. Experience indicates that comminutors suffer of numerous operational problems.

Grit chambers. Grit removal is an important treatment to reduce abrasion and wear of mechanical equipment, grit deposition on pipes and channels. There are 5 categories of grit removal: aerated chambers, vortex-type, detritus tank and hydrocyclone units and horizontal grit chambers. Only the horizontal grit chambers are used in Latin America. Grit disposal requires care due to its organic content. Usual disposal point is a landfill. Typical design criteria is shown in the following table.

Description	Range
Horizontal velocity in m/sec	0.15 to 0.4
Detention time at peak flow in seconds	15 to 90
Length in m	3 to 25
Water depth in m	0.6 to 1.5

Description	Range
Velocity control	Proportional weirs

Scum removal. The scum units are used to eliminate oil, grease, plastics and other floating materials.

Primary treatment typically refers generally to sedimentation.

Sedimentation. The theory, operation and parameters of the sedimentation units have been discussed on the previous water supply section. In regard to the overflow rates it varies from 12 to 40 $m^3/m^2/day$. The detention time varies from 1 to 4 hours. Sedimentation can be enhanced by the chemical applications. Criteria for rapid mix, coagulation and flocculation are similar to those already discussed in the previous water supply section. Settled primary sludge is typically scraped into a hopper from which it can be disposed by gravity or by pumping. For scraping a traveling bridge is commonly used. Tanks can be rectangular, circular or square. When deemed appropriate settling tanks can be designed as stacked (2 units in one) or using inclined settling plates. Cleaning the plates is more difficult for wastewater treatment than for water. In primary sedimentation the efficiency in Suspended Solids removal varies from 50 to 65% and for BOD_5 removal, 30% is a common value. For enhanced sedimentation SS efficiency can increase to 75%. To complete primary treatment it is necessary to digest the sludge either by anaerobic or aerobic processes. Digestion is the most difficult and expensive component of primary treatment. Anaerobic digestion is the more cost effective process.

Imhoff tanks. Imhoff tanks are an old and very appropriate technology for primary treatment, and in Latin America, the Imhoff tank process is still popular. An Imhoff tank combines in one single unit sedimentation on top and anaerobic digestion at bottom. Its operation is simple and the process is low cost. The stabilized solids from digestion can be disposed safely as compost. The surface loading rate varies from 32 to 40 $m^3/m^2/day$; detention time varies from 2 to 4 hours. In Latin America this technology has been adopted in many cases due to its operational and cost advantages. To facilitate its operation the flow of the tank can be reversed. See reference (18) for information about capital costs and operational costs.

Secondary treatment or biological treatment addresses the need to reduce the BOD_5 and SS by about 85%. There are several unit processes in this treatment category. The common classification is:

- Suspended Growth Biological Treatment that includes: conventional activated sludge, contact stabilization, extended aeration, and oxidation ditches;
- Attached Growth Biological Treatment that includes: trickling filters, rotating biological contactors, and up flow anaerobic sludge blanket (UASB).
- Stabilization ponds are also biological treatment, and are currently classified as part of the natural systems category that is briefly described below.

Both suspended and attached growth secondary treatment systems are possible.

Suspended Growth Biological Treatment commonly includes the conventional activated sludge, contact stabilization, extended aeration, and oxidation ditches.

Conventional Activated Sludge. This is the most commonly used treatment in the USA since 1940. For the purpose of this presentation, a brief description follows. (See ref 29 for more details about design and operation). The train of treatment of a conventional activated sludge includes, after the primary settling tank, an aeration tank (contact reactor) followed by the secondary settling tank. Wastewater biological solids are first combined, mixed and aerated in the reactor for about 6 hours. Content of the reactor (mixed liquor) consists of microorganisms, inert material, biodegradable and non biodegradable materials, SS and colloidal matter. The particulate fraction is called the mixed liquor suspended solids (MLSS). After sufficient time in the reactor, mixed liquor is transferred to the secondary settling tank to permit separation of the MLSS by gravity. A portion of settled MLSS is returned to the reactor to maintain a concentrated microbial population at the rate of 25 to 50 % of the influent rate. The excess quantity of sludge is disposed. The conventional sludge treatment can operate by tapered aeration, continuous stirred, or step aeration. Many variations and combinations are possible.

Contact stabilization. This is also an activated sludge process with some modifications to the conventional process. In some cases the primary settling tank may be eliminated. A contact tank is added after the reactor in order to provide additional aeration. This process modification may be operated at a higher flow-through capacity than conventional activated sludge, but is more difficult to operate effectively.

Extended aeration. This process requires a long aeration time and is designed for lower loading rates. Primary sedimentation is omitted from the process to simplify the sludge treatment and disposal. Aerobic digestion of the excess solids followed by dewatering on sand beds is often incorporated.

Oxidation ditch. This process has been developed in Netherlands for small towns and it is in essence an extended aeration process. It is also called carrousel process. The channel conveying the wastewater is about 1.5 m wide and 1.5 m deep, has one or more rotors for mixing and oxygen addition, and a waste sludge disposal line. In Latin America there are many applications of this process due to its simplicity and low operation cost.

Attached Growth commonly includes the trickling filter, rotating biological contactor, and the sludge blanket.

Trickling filters. Trickling filters have a filter medium, usually formed by rocks or plastic through which the wastewater is percolated. In essence, in every opening of the trickling material, a mini sedimentation unit is formed and the biological content of the wastewater is attached and treated on the surfaces of the medium. Filters have an underdrain system to conduct the effluent away. Natural or forced aeration is needed to help the trickling process aerobic. The size of the rocks varies from 1 to 4 inches. The depth varies from 1 to 2.5 m. Usually a primary settling tank or Imhoff tank is located ahead of the filter. The trickling filters can be of low rate (1 to 4 m^3/m^2/day) or of high rate (20 to 25 m^3/m^2/day). For the high rate trickling filters a recirculation line is provided. In the opinion of the author, the combination of an Imhoff tank and a trickling filter is appropriate technology that should be used in Latin America.

Rotating Biological Contactors. Rotating biological contactors are manufactured as package plants. RBC is a fixed film biological reactor using a series of rotating plastic plates, mounted on a horizontal shaft and partially immersed (about 40%) into the wastewater. The discs are slowly rotated while wastewater flows through the tank. Rotation results in exposure of the film to the atmosphere as a means for aeration and oxidizing of the organic matter. Currently a pilot plant is needed to obtain the proper parameters for design. Pilot plant study is recommended for all plant design projects, especially when industrial wastes are a component of the wastewater flow.

Up Flow Anaerobic Sludge Blanket. This process combines in a single reactor mixing, coagulation, liquid solids separation and sludge removal. This unit can be classified as a filter[(18)]. In Brazil this technology is extensively used successfully and is referred to as RAFA (reactor anaerobio de flujo ascendente).

Stabilization ponds are appropriate technologies for use in Latin America. CEPIS has conducted numerous studies and has published their results in several publications. There are three types of stabilization ponds (also called oxidation ponds), defined according to the biological activity taking place: aerobic, anaerobic, facultative.

Aerobic ponds. In the USA aerobic ponds are used primarily for the treatment of soluble organic wastes and effluents from wastewater treatment plants. Due to weather conditions in the USA, aerobic ponds for untreated wastewater are designed for very conservative BOD_5 loadings, in the range of 100 kg/hectare/day. In Latin America where the climatic conditions are warmer the common loading is about 300 kg/hectare/day. Aerobic stabilization ponds are shallow earthen basins of varying geometry, involving the use of algae that produce oxygen during photosynthesis, sometimes using mechanical aeration on the surface, more often depending on the wind action. Aerobic conditions should prevail throughout the depth. These units require a long detention time of about 7 to 30 days. Depth is about 1.0 to 1.5m. The inlet of the pond is located at the center. They do required space below for sludge accumulation. Usually some cleaning is done every 1 to 2 years. The efficiency of aerobic ponds in terms of BOD_5, varies from 60 to 75% removal. Ponds can be operated in parallel or in series depending on the desired output. One of the main features of the pond is its high efficiency to remove bacteria. In the experimental pond system built several years ago in San Juan, an area near Lima, Peru, several tests were carried out to obtain design parameters. Aerobic ponds working in series were capable of 99.99% reduction of fecal coliform. Because of the large area required, this technology is most useful when cheap land is available.

Anaerobic ponds. These units operate under anaerobic conditions primarily to treat high strength organic municipal or industrial wastewater that contains a high concentration of solids. Commonly, an anaerobic pond is an earthen basin about 6 m depth. Its size is based on surface loading rates that are higher than those for aerobic ponds. Depending on the operating conditions the anaerobic ponds are efficient to between 70 and 85% BOD_5 removal. The detention time is about 20 to 50 days. The required area may be less than for aerobic ponds due to the use of higher loading rates.

Facultative ponds. Typically, wastewater to facultative ponds receives no more pretreatment than screens. These units have a surface layer where aerobic conditions prevail, an intermediate layer where facultative bacterial action is partly aerobic and partly anaerobic, and an anaerobic layer at the bottom where only anaerobic conditions prevail. Commonly these ponds are designed to obtain a BOD_5 effluent of 30 to 40 ppm, and SS of 40 to 100 JTU. Facultative ponds are usually designed in series. Due to the mixed operation it is difficult to predict safe loads for design. Parameters may be obtained from a pilot plant.

Aerated ponds. When available area is a limiting design factor, aerated ponds are used. Detention time varies from 3 to 10 days. These units require the introduction of oxygen to accelerate the stabilization process. Air can be provided by floating surface units or by diffusion lines located at the bottom. In Latin America the operation of mechanical equipment is difficult due to the intermittent failure of power sources as well as the lack of trained personnel. To assure the efficiency of this process precise control of SS is required.

Land Treatment (Natural Systems)

There are three types of land treatment: slow rate, overland flow and rapid infiltration. In slow rate and rapid infiltration, treatment is achieved during percolation of the wastewater into the soil. In the overland type the treatment occurs in a sloped thin layer of soil. The following table prepared by US EPA presents the minimum pretreatment prior to land application.

Land treatment system	Requirement
Slow rate	Primary treatment is acceptable in isolated areas. Ponds are acceptable when the effluent is no more than 1000 MPN/100ml of fecal coliform. To apply the effluent on parks and golf courses the minimum is 200 MPN/100ml of fecal coliform.
Overland flow	Screening or comminution is acceptable for isolated areas. Screening or communition plus aeration to control odors is acceptable with no public access.

Land treatment system	Requirement
Rapid infiltration	Primary treatment is acceptable for isolated areas. Ponds are acceptable for urban areas with controlled public access.

Tertiary treatment. Tertiary treatment or advanced treatment, addresses the need to reduce, in addi Tertiary Tertiary tion to the BOD_5 and SS, phosphorous, nitrogen, metals, refractory organics and other substances that may be discharged to the environment, particularly in developed countries where the industry is ever producing new and more refractory elements. In Latin America this kind of contamination is limited to locations with industrial wastewater.

After the treatment processes reviewed above, it is relevant to indicate the status of the technology development in the USA and in Latin America. In ref 18 the author presents the review of 44 current technologies and 26 case studies related to the existing and new technologies in the USA. The review also presents information about design parameters, cost, advantages and disadvantages. It is a very useful reference that should be consulted.

In Latin America, except for Imhoff tanks, trickling filters, stabilization ponds, based on investigations conducted by CEPIS, and the RAFAS process, no investigations or research studies have been carried out related to the use of imported technology. The current coverage of wastewater treatment in Latin America for domestic wastes is very limited, in the range of 3 to 5% (see chapter I). From an environmental perspective it is expected that treatment technology applications in future projects will assure the minimization of adverse environmental impacts. Hence, more research and technology application studies should be developed to obtain an appropriate regional technology.

Freshwater and Marine Outfalls. Outfalls are an efficient wastewater disposal process for communities located on the coast. Ocean disposal must be distinguished from disposal to estuaries and embayments because of limitations on flow of the combined wastewater/seawater mix in the latter cases. As an example of efficiency, particularly in the reduction of fecal coliform, the outfall with pretreatment designed for a wastewater project in Montevideo, Uruguay a few years ago has the following results: initial dilution (1/85), transport dilution (1/1.8) and die-away (1/657). Combining these 3 factors the reduction is as high as 5 orders of magnitude. Chemical disinfection would be required to reach so high a reduction value.

Notwithstanding, in a recent project in Panama it was not found to be acceptable. The local environmental regulations required biological treatment as the minimum acceptable treatment before final disposal.

Wastewater Reuse. Reuse of treated wastewater effluent by direct or indirect modes is a recognized disposal method. Professor Shuval, from the University of Israel, in a conference on reuse held in Washington D.C. several years ago, said "the time will come soon when a person looking at a drop of municipal water says, 'I know you' ". Today, water used as domestic water supply on most rivers on the planet has already been used several times before the last downstream user drinks it. There are many examples of wastewater reuse projects all over the world. Fresh water is recognized as the limiting natural resource on the planet and the situation will become more pressing as the population grows. In San Diego, California an experimental plant is reusing a small flow of the collection system for drinking purposes after the wastewater receives additional treatment including the use of hyacinths. In an area South of Lima, Peru, an indirect wastewater reuse project is in place using the effluent of a pond system to irrigate some crops that are not consumed raw. In Israel, there is a famous project called the Dan Project where the wastewater, after advanced chemical treatment is used to recharge the groundwater aquifer for drinking purpose. The table below[29] indicates the common uses of reused wastewater.

Use	Direct	Indirect
Municipal	Park, golf and lawn watering. Potential for municipal water supply	Groundwater recharge
Industrial	Cooling water, boiler feed water and process water	Groundwater recharge for industrial use
Agricultural	Irrigation of certain agricultural lands, crops, orchards, and forest; leaching of soils	Groundwater recharge for agriculture
Recreational	Artificial lake for boating, swimming.	Develop fish and waterfowl areas
Other	Groundwater recharge to control saltwater intrusion; wetting agent for solid waste compaction	Groundwater recharge to control land subsidence, soil compaction

Water and Wastewater for Rural Areas. IADB has a wide experience in the development of water projects to serve rural areas of Latin America. IADB was the pioneer international lending organization committed to supporting rural programs. In the 1960's and 70's its lending represented almost all the international funds allocated to the region. Since then, in most of countries rural water programs have been developed by stages (some countries are in the 6th project stage), like in Argentina. The design of these programs, called "multiple works", is based on the appraisal of a representative sample of about 30% of the total cost of the program. The sample is appraised to determine the soundness of the projects and their investment costs. Based on this information the total cost of the program is estimated. Every Bank project to be included in the program must have a positive Net Present Value (NPV) using a discount rate of 12%.

The appropriate wastewater disposal method for rural areas is in-situ disposal (dry or wet units). Every rural water program should include a very simple disposal unit (latrine), as a minimum.

Water Quality of Receiving Waters. It is usual to define the water quality of the receiving waters based on its beneficial use. It is common in water resource management legislation to classify the quality of the receiving waters by use. A typical classification and limits of the critical components for inland waters and for marine waters are presented in the following Tables. Almost all the countries in Latin America have similar classifications.

Inland Waters

Class	Description	Dissolved Oxygen	Floating solids, oil grease and scum	Color and turbidity	Coliforms	Taste and odors	pH
I	Water supply	As naturally occurs	None others than natural origin	None other than natural origin	MPN 100/100 ml.	None than natural origin	As naturally occurs
II	Bathing and other recreational activities	Minimum 5.0 ppm	None	None	MPN 1000/100 ml	None or in such concentration that impairs taste and odor in edible fish	6.5 to 8.0
III	Fish, wildlife and fishing	Minimum 5.0 ppm	None	None	MPN 1000/100 ml to 2300/100 ml	None or in such concentration that impairs taste and odor in edible fish	6.0 to 8.0

Class	Description	Dissolved Oxygen	Floating solids, oil grease and scum	Color and turbidity	Coliforms	Taste and odors	pH
IV	Navigation	2.0 ppm	None	None	Same as above	None or in such concentration that impairs usage	6.0 to 9.0

Marine Waters

Class	Description	Dissolved Oxygen	Floating solids, oil grease and scum	Color and turbidity	Coliforms	Taste and odors	pH
I	Suitable for all seawater uses	Not less than 5.0 ppm	None	None or in such concentration that impairs any usage	MPN 70 to 230/100 ml	None	6.8 to 8.5
II	Bathing and other recreational activities	Not less than 5.0 ppm	None	None or in such concentration that impairs its use	MPN 700 to 2300/100 ml	None or in such concentration that impairs its use	6.8 to 8.5
III	Fish, shellfish and wildlife	Not less than 5.0 ppm	None except that amount resulting from discharge of WWTP	Same as above	Same as above	Same as above	6.8 to 8.5
IV	Navigation	Not less than 2.0 ppm	Same as above	Same as above	Same as above	Same as above	6.0 to 9.0

For a water pollution control program to have real impact on the quality of the receiving waters, all significant discharges to the water body must be controlled. In the old legislation for receiving waters it was usual to include the degree of treatment. This approach was considered as "end of pipe treatment" for point discharges. Non-point discharges or diffuse sources of pollution were not considered. Construction of isolated treatment plants alone will not provide the expected control benefits for receiving water quality, nor will the investments in plants that are later non-operational. For these reasons an integrated approach to pollution control is needed, in which all the elements of a control program are in place including: area wide planning, regulations, monitoring and enforcement, plant design, construction and operation, financing, cost recovery, institutional coordination and technical assistance for training, information systems, research and public education.

In Britain, since 1966 the Royal Commission's general standard as a norm for wastewater effluents is: BOD_5 20 ppm and SS 35 ppm, with a dilution factor of 9 to 150 volumes in the receiving watercourse (the watercourse having not more than 4.0 ppm BOD_5). In the USA the minimum treatment is secondary treatment. The minimum national performance standards established by EPA are indicated in the following table.

Parameter	30-day average	7-day average
Conventional treatment		
1. BOD_5		
a. effluent ppm	30	45
b. percent removal	85	
2. Suspended Solids		
a. effluent ppm	30	45
b. percent removal	85	
3. pH	6.0 to 9.0	
4. fecal coliform MPN/100 ml	200	400
Trickling filters and stabilization ponds		
1. BOD_5		
a. effluent ppm	45	65
b. percent removal	65	
2. Suspended Solids		
a. effluent ppm	45	65
b. percent removal	65	
3. pH	6.0 to 9.0	
4. fecal coliform MPN/100 ml	200	400

Water Quality Modeling. Most of the information of this section comes from references (30) and (31). To facilitate the integrated or holistic approach mentioned above, water quality models have been developed that attempt to simulate changes in the concentration of pollutants as they move through the environment. The main purpose of most water quality models is to determine the new concentration of a parameter at specific points in the environment when the modifying conditions and existing situations are known.

A model is reliable when it can predict future concentrations. There are two steps to developing models, verification and validation. Verification is the step that determines that the model is doing what it was designed to do. Validation is the step that determines that the model predicts behavior in the environment. Many models are used actively that have never been validated, and are nevertheless found to be acceptable, such as several air quality models. Whenever possible, it is essential to test the output of any model against observations for several water flows and concentrations.

When a contaminant is introduced to a body of water it undergoes three fundamental actions: dispersion, transport and degradation. Dispersion or diffusion is a function of the concentration gradient. This can occur at the molecular level due to the Brownian motion causing random movements, it is also due to turbulent eddies and velocity shear. Advection refers to transport due to the bulk movement of the water that contains the contaminant. Decay is the transformation of the contaminant due to physical, chemical and biological reactions or a combination of all of them. The output might be an increment or decrement of the concentration of the contaminant, and it is independent from the other two causes. As examples of the application of the water quality models two cases are presented below.

Case 1[(32)]**.** Panama City Wastewater Master Plan and Feasibility Study. Panama's metropolitan area has twenty one political districts spread over two watersheds located to the east and to the west of the Panama Canal. The study area consists of about 35,000 hectares. Only two districts are located to the west of the Panama Canal. For both districts, independent wastewater systems are designed and consequently, they were not included in the alternative study of the Master Plan. The slope of the area is from north to south toward the bay. There are eight rivers crossing the eastern part of the area discharging to the bay. Only five rivers were relevant to the study. The current population (including equivalent population) is about 0.8 million discharging a mean flow of about 4 m^3/sec. The estimated population (including equivalent population) for year 2010 is about 1 million discharging a mean flow of about 5.1 m^3/sec. The estimated population for the year 2020 (end of the design period) is 1.2 million discharging about 6.2 m^3/sec.

The Panama City metropolitan area, including the bay, has the following environmental problems:

a. Some populated areas are without sewerage service.
b. Rivers are contaminated by wastewater discharges as well as by solid waste.
c. The existing collection system has limited capacity and produces flooding.
d. Water courses are contaminated by industrial discharges.
e. The bay is contaminated by more than 30 raw wastewater discharges.
f. The bay is contaminated by liquid wastes that are disposed by a daily average of 36 ships.

The aim of the Master Plan is the design of structural and non structural measures to clean up the metropolitan area and the bay. The bay water quality is not sufficient for bathing but is used for fishing. Once the Master Plan is implemented the bay can be safely used for recreational activities.

The Master Plan provides an adequate solution using an integrated approach to all the present contamination in the study area. The Plan is the result of a comparison of 18 alternatives at the prefeasibility level. The selected alternative has a positive present value and is designed in two stages and consists, in addition to the collection and transportation system, one sea outfall of 6.5 km with preliminary treatment, microscreen and disinfection, and four secondary

WWTPs with disinfection to intercept the wastewater flows discharging to the rivers. The interceptor design uses two collectors along the rivers connecting to a single WWTP. The typical design for the secondary plants include: one upflow anaerobic sludge reactor (RAFA), one facultative lagoon and sludge drying beds. The effluents of the secondary treatment plants have the following concentrations: BOD_5; 35 ppm and SS; 35 ppm. These values are considered adequate to protect the water quality of the rivers and the bay. The first stage has an estimated cost of about US$ 215 millions.

For the preparation of this study three mathematical models were used as described below:

1. The first one named SWMM has been developed by CaiCE software corporation of Tampa, Florida USA. It is used to design the new collection system as well as the rehabilitation of the existing system. It is supported by GEOPAK for the location of the treatment plants, and Microstation that is a comprehensive package for CAAD.
2. The second model is the QUAL2E developed by USEPA, and it is used to simulate the water quality of the 5 principal rivers. Using this model it is possible to determine:
   a. the water quality of the rivers along its courses, to verify the compliance with the quality standard adopted, once the project is implemented.
   b. the level of wastewater treatment necessary, to maintain the water quality of the rivers as required by the adopted standard.
   c. make recommendations for the location of the WWTP.
3. The third one is a package of 2 models: Delf3D to analyze the hydrodynamic behavior of the bay (impact of winds and trend of the water movement), and Delft 3D Waq to analyze the water quality and the morphology (benthos sediments). Both models were developed by Delft Hydraulics Laboratory. The following parameters were incorporated in the models: SS (inorganics), SS (organics), BOD_5, Oxygen, Ammonia, and fecal coliforms.

These models were used for the following applications:

1. design of the sea outfall and analysis of the sediments.
2. determination of the impact of the final discharges from the rivers on the water quality of the bay.
3. Determination of the fecal coliform contamination along the shore line of the bay; to assure that the MPN 2000 fecal/100 ml, established as a limit, is not exceeded.

The non structural recommendations of the study are to:

1. design and implement the water quality standards for receiving waters at a national level;
2. enforce the current legislation to control the effluent discharges to the wastewater system, bodies of water and the bay;

3. monitor the degree of contamination of the rivers;
4. enforce the current legislation to control the solid waste discharges to the wastewater system as well as to the rivers;
5. enforce the current legislation to control the wastewater discharges from ships to the bay.
6. implement the environmental plan;
7. enforce the water quality monitoring requirements;
8. develop a training program for the responsible unit of the government IDAAN (Instituto de Agua y Alcantarillado Nacional);
9. develop a public education program.

Case 2[33]. Tietê River Pollution Problem. The parts of the Tietê River and its tributaries that flow through the São Paulo Metropolitan Area (SPMA) in Brazil, are the most polluted bodies of water in the state. The Tietê River enters the metropolitan area with acceptable water characteristics but in Guarulhos, at the confluence with Jacu, it becomes anaerobic. Downstream from Jacu, the large volume of domestic and industrial waste dumped into a relative small flow has made the river an open sewer that supports no aquatic life, smells most of the year, and is used only as a sewer canal for more than 80 Km.

In 1985 the State sanitation company (SABESP), began to address two problems: a) low sewerage coverage (64%) in its metropolitan service area, and, b) river pollution caused by not treating 81% of the sewage that was collected. The Company contracted for a Master Plan to study the least costly way to clean up the river. The study focused on the Tietê basin within the SPMA rather than the whole basin. This was a reasonable simplification because the river is 1000 km long.

The impact on river quality caused by reducing the inflow of contaminants at different points was simulated using QUAL2E model developed by USEPA. The model is deterministic and relatively simple. Hydrological variations are determined external to the quality model and the quality is calculated on the basis of a particular river flow and the contaminant loads that enter. The model separately accounts for contaminant loads coming in above the SPMA, point discharges of industrial wastewater, sewerage outfalls, contaminant loads from tributaries, non point discharges from surface runnoff, and river inflows from groundwater. The construction cost estimated by the Master Plan including connections, interceptors, treatment plants and disposal of solid waste, was estimated at US$ 1.45 billion in three stages. The water quality in the Tietê Basin at minimum flow at the conclusion of every stage is expected to be as indicated in the table below.

Point Where Quality Is Measured	Dissolved Oxygen (ppm)	BOD_5 (ppm)
Tietê at Tamanduateí		
1998	0	33.49
2001	0	23.19
2010	1.46	13.64

Point Where Quality Is Measured	Dissolved Oxygen (ppm)	BOD_5 (ppm)
Tietê at Pinheiros		
1998	0	15.22
2001	1.33	3.73
2010	1.14	3.70
Pinheiros pumping station		
1998	0	18.47
2001	1.99	7.21
2010	2.07	7.2
Edgard de Souza Dam		
1998	0.29	29.16
2001	1.34	22.88
2010	2.48	12.65
Pirapora		
1998	2.24	20.6
2001	2.6	17.59
2010	3.03	10.97

ENVIRONMENTAL EVALUATION

Methodology(11)(26)

For all the water and wastewater projects falling in categories III and IV of the IADB's nomenclature, an Environmental Impact Assessment (EIA) is required. The EIA should be prepared along with the study of alternatives.

The preparation of an EIA consists of numerous activities, investigations and technical tasks to identify the main environmental consequences of the project comparing the situation "with and without" the project. This methodology should not be confused with the situation "before and after" a project.

The activities mentioned are: project description, legislation, environmental diagnostic, identification and analysis of the potential environmental impacts, mitigation measures, monitoring and a final report.

For the execution of the various tasks several methods and techniques called "instruments" are used to identify, appraise and summarize the environmental impacts. Some of these instruments are used in the different disciplines involved and others allow the integral and multidisciplinary approach required by the EIAs.

Among several methodologies the more common are:

1. Ad hoc methodology, that consists of the evaluation of the project and the identification of the impacts made by a group of experts. Generally is used when the timeframe is short. This methodology is very subjective.
2. Check list methodology, that consists of the preparation of a list of all the environmental impacts generated by the project. Check lists are currently in use in the environmental diagnostic and on the comparison of alternatives. The check list numeric values are adopted to indicate the intensity of the impacts varying from 1 to 3, 3 being the most intensive impact. This is a much better methodology that permits establishment of the cause-effect relationship.
3. Interaction matrix is a more complete methodology in which on the "X" axis the list of impact factors are included and in the "Y" axis the list of activities (components) of the project are included for the construction phase and for the operational phase. This methodology was developed in 1971, by professor Leopold *et al.* The Leopold spreadsheet consists of a list of 100 environmental impacts and a list of 90 activities resulting in 9000 cells in each of which the intensity of the impact is included in a range from 1 to 10, 10 being the more intense. The analysis is subjective. Once the impact is identified a diagonal line is written on the cell. At the end of the exercise the list of impacts and their intensity can be clearly seen. Also the magnitude of the impact and their importance is presented. The IADB's appraisal methodology for investment projects requires that all environmental impacts are quantified and included in the cost benefit analysis.

Typical Simplified Content of EIA Report

1. Purpose of the report
2. Background
3. Project area (setting) described in detail:
   a. Physical setting: topography, geomorphology, geology, rivers, lakes, hydrographic basins, ruins and archeological areas, fishing, water quality, air quality, sounds, hazardous residues.
   b. Biological settings: fragile ecosystems, aquatic plants, animal habitat, migration, natural reserves, population within the area, flora and fauna, flooding areas, swapping areas, species in danger.

c. Human resources settings: demography, economy and development, lands use, infrastructure, housing, schools, health care, transportation, recreation, public services, archeological and historic resources.

4. Description of the environmental legislation and characteristics of the local responsible entity.
5. Description of the most relevant parts of the alternative study in a comprehensive manner to permit the reader a clear understanding of the project without consulting additional documentation.
6. Description of the environmental impacts: positive and negative; reversible and irreversible; permanent or temporal; environmental impacts during construction phase.
7. Quantification of the environmental impacts to be included in the socioeconomic evaluation.
8. Identification and quantification of the mitigation measures for negative environmental impacts.
9. Description of the environmental Plan including the methodology for its implementation, the schedule, the monitoring and human and economic resources required to comply with. Special attention will be given to the relocation of the existing population to minimize their disruptions. The migration should be carefully organized and the affected population should receive a fair compensation. In some projects, the relocation of the population has been the critical task.
10. Community relationship is very important for the success of the project. Seminars and public presentations should be held along the preparation of the report and their results should be made public.
11. Final Report and Executive Summary. Special care should be taken in the preparation of the Executive Summary because it is the first document reaching the decision- making levels. In IADB is usual to submit this report to the Board of Directors ahead of the loan proposal. See Appendix III.

Institutional Evaluation

Since ancient times there has always been a responsible entity (some times a single person) to manage the water delivered. Starting from simple rural towns to mega cities, an entity for planning, managing, operating and regulating the systems is indispensable.

The old Agenda, focused on the incremental improvement of household services without caring about the environment posed large financial, technical and institutional challenges. Currently the central authorities are responsible for carrying out the three functions using the revenues from the water rate and some subsidy from the central government.

This scheme was plagued with unskilled people sometimes hired to pay political favors, great inefficiency and bureaucracy. Today in some countries in the Caribbean, e.g., Jamaica and Bahamas, this type of management still exists and the quality of the service may be considered fair. Once the new trend of private sector participation is suggested to them as a new model, they react adopting the sub model of outsourcing some functions like meter readings, billings, new in-house connections, maintenance and operation of pumping stations and water and wastewater plants, and other functions. There is always reluctant to reduce manpower in order to avoid the social issue. Waiting for retirement age makes reduction subtle.

In some large cities in South America the private sector participation model is applied with relative success. The difference now is the creation of autonomous regulatory entities that control government owned institutions. Gaining approval of new water rates is still a troublesome social issue.

In other large and medium size cities the responsibility for management of the Water and Wastewater systems falls under special departments of the municipalities. This arrangement, that has proven to be successful in Central America, depends on the municipal economic resources being separable from the general revenues of the municipality. Generally speaking the municipal owned water services make a great contribution but the political influence is still there. Recently, the trend is to create autonomous municipally owned entities with a mix of private and municipal capital.

In Argentina, the participation of the private sector using concession type contracts was adopted as a model for the entire country with some difficulties. The main problems arose from the concessionaire trying to increase water rates, and from the government, insisting on compliance with the planned investment program by the concessionaire. Notwithstanding, in a recent declaration of AIDIS-Argentina chapter, it is requested that the authorities at different levels define new sector policies, to plan and coordinate all the state actions, to increase investments (local and international) and to establish adequate regulation and control at the country level. It seems that the old decision to privatize the water utilities in Argentina is under review now.

In Chile, the concessions were adopted by a combination of private (international and local) and government entities. The government owns only 35% of the total shares of the corporation.

For rural towns the creation of simple cooperatives has been the rule and also has been successful. In effect, the main problem of this model has been, in the long run, the poor maintenance and the lack of sufficient economic savings for expansion of the systems. It is very common that governments request international funds to rehabilitate poorly kept systems.

As described in this general discussion there is not a standard model to be adopted in all cases. Hence, the model has to be properly adapted to a specific situation(s) to guarantee the sustainability and the accountability of the services. But the most important feature in all models is the need for political will and active public participation. Besides, in the New Agenda of environmentally sustainable development that has emerged after the United Nations Río de Janeiro's conference (Agenda 21, held in June 1992), much greater attention should be placed to ensure that our use of water resources is sustainable both in terms of quality and quantity. To respond to these new challenges the institutional approach has to be innovative.

FINANCIAL EVALUATION

The financial analysis or management assessment of a public water utility, (historic analysis versus economic analysis), is a time consuming and complex activity that generally speaking, is out of the scope of this book. However, some guidelines of the main tasks are presented here. The financial management must focus on: a) Planning (short, medium and long range), and Budgeting, b) Revenues and Collection methodologies, c) Procurement procedures and standards, d) Accounting system including cost, financial and managerial accounts, and existing software capabilities e) Cash generation and Debt management including national and international obligations, f) Internal and external Auditing, g) Information system to produce timely reports for all the organization.

There are some financial ratios that help quickly to understand the financial standing of the utility. Among them are the following: a) rate of return, b) assets turnover, c) account receivable turnover, d) current (current assets over current liabilities, e) quick or acid test (total liquid assets over total current liabilities), and f) debt- service (net income before interest and depreciation over total debt).

In the end, the purpose of the financial evaluation is to verify that the project is feasible from a financial standpoint and to define the long term planning and financial strategy of the entity. In other words that the revenue from the water rates and other charges are sufficient to support the operation, maintenance, depreciation and debt service of the institution, for present and future situations. In addition, the water rates should generate internal funds to support future expansion of the system. In some cases, when the responsible entity is financially weak the government assumes all the debts of the enterprise. Consequently, the enterprise is free of debts and able to enter into new commitments. Cost recovery mechanisms should be sufficient at least to support operation and maintenance of the system, as a threshold.

An early financial analysis is made using the current water rate to determine the gaps and the needs for an increase in water rates to carry out the expansion project. Once the need for an increase in the water rates is indicated, rates may be raised gradually to facilitate the understanding and cooperation of the users. A well conceived Public relations program, is crucial to attain this objective. It is customary to make a financial flow analysis for the first 10 years only. IADB uses a digital model SPIMOD to facilitate its work analyzing different financial scenarios.

Some consulting firms have developed proprietary digital models to facilitate the financial analysis of projects and to demonstrate financial sustainability in the long run. These models are made with ample capacity to analyze different investment scenarios too. The total investment required is estimated using technical and cost information about the project or program for a 20 year horizon. The model must permit projection of the financial situation of the entity taking into account all variables like revenues, costs, investments etc. Indicators such as the Internal Rate of Return (IRR) and the NPV may be developed quickly for each of the alternatives and scenarios. Appropriate digital models are useful tools and its use should be encouraged in all Latin America utilities.

ECONOMIC EVALUATION

Methodology for the Estimation of Economic Benefits

Ideally, the benefits of water and wastewater systems should be measured together as a single service. Benefits are to be measured by the consumers' willingness to pay as defined by the area under the relevant portions of the demand curve for the services provided.

The demand curve for WAW services may be estimated econometrically using the following functional form:

$$Q = f(P, Y, N, e)$$

Where:

Q = consumption of water
P = incremental cost of consuming water
Y = monthly family income
N = family size
e = statistical error term.

This function is usually estimated from time series data. In most cases however reliable estimates can be obtained from cross sectional sample survey data. To obtain workable coefficients in the above functional form it is important that a wide range of observations, especially for price and income be included in the sample. For this reason, it is useful to ensure an adequate representation in the sample of people without service or with alternative high cost services. A broad range of income levels should also be represented in the sample.

The functional forms that give best results are usually "constant elasticity demand curves," although linear demand curves may also be tested. The IADB's SIMOP software package may be used to perform the cost-benefit calculations for water projects. This program has been available since 1978 and has been applied in numerous water projects.

There are some cases, where pricing data do not support an analysis of potable water and wastewater as a joint service. In such cases, the demand analysis described above should be done separately for potable water and alternative methods might be needed to estimate the benefits of expanding wastewater service, as set forth below.

Economic Model for Water Projects

The model purports that, among other factors, there is a negative relationship between quantities consumed and prices charged, and a positive correlation with income levels and family size. Also for an average sized family with a particular level of income, one can use the demand function to plot how quantities of water consumed would decrease if prices charged were to be increased.

This relationship is of extreme importance in the analysis of water projects since every point on the curve expresses the willingness to pay (WTP) of users for the last liter of water being demanded. Total WTP, given by the area under the demand curve for prices in the "with" and "without" project situation is taken as an approximation of the value in terms of monetary income that the user assigns to the improvement in the "with" project case. Demand functions should be central to the efficient management of water supply systems, both from a day to day financial and operational point of view as well as in planning the expansion of a water system.

The process starts when the planner sets initial prices for water, usually those at the current level. Population and income level projections are exogenous inputs which together with the price level provide the necessary information to be plugged into the demand function to arrive at consumption levels for each year of the planning period. Industrial, commercial, public water use and water losses are then added to complete the projections of demand of the system.

The estimation of demand curves for water requires a wide array of observations of water behavior from different types of consumers at different price levels. This has been done successfully in developed countries using billing data from the utilities but in Latin America and the Caribbean, this source of information by itself is not, in general, very useful since very little variance exists in water prices charged by the water utilities. In fact, marginal water prices are generally subsidized at low levels and are zero in many cases because most of the systems are unmetered.

In order to find more variance in prices charged, the market has to be searched specifically for family units who are not being served by the water utility and therefore have to pay higher prices, to other suppliers. In the upper price range we can find water users who have to resort to water vendors, or spend many hours per day hauling water from public taps or sources that often lie distant from their homes. In the mid price range we obtain information from water users who have irregular service from the public utility and even though enjoying indoor plumbing and appliances, have to purchase water from private sources who deliver by truck. Generally, such users own cisterns with enough capacity to fulfill family needs for periods of more than one week. Also in this price range we find households with private wells who have to pay high variable pumping costs in addition to other fixed costs (wells, pumps, storage tanks, etc.).

In the lower price range we use information obtained from the water utilities. From billing data we select only those customers that have 24 hour service, and are charged for each additional unit of water. A random sample of these accounts is selected upon which, field work is performed in order to inquire about income and family size.

Information on income, family size, quantities and cost (monetary or otherwise) for users paying mid and upper prices for water is gathered with household surveys. The last part of a typical survey questionnaire is dedicated to ascertain family income and other socioeconomic information. Apart from it being a very private matter, problems arise from the fact that income can be perceived as affecting taxes or utility pricing. Also, inaccurate information results from incomplete answers, as the respondent excludes or forgets income derived from all the household members. In order to avoid these issues it becomes necessary to list all members of the family and inquire separately about each of their income.

The first section of the questionnaires is generally dedicated to obtain information about water use habits and prices being paid. Since one of the main objectives of the survey is to get data on high prices for water we refer the respondent to the sources used at times of scarcity. Similar questions are asked in surveys to be carried out in urban settings, where we target population located in areas suffering from water scarcity, and therefore alternative water sources are used like, bottled water, public standpipes, neighbors, private wells, water vendors, etc.

Both linear and logarithmic functional forms are experimented with during the estimation process. The form finally chosen to estimate the WTP is the one capturing the greater variance, and provides the best fit. In general, linear forms seem to provide better fit when many observations contain near zero marginal prices.

Economic Model for Wastewater Projects

The expansion of sewerage services (including collection, transmission, treatment, and final disposal) may entail two benefits to those who receive the service:

1. New customers of sewerage service may save some money because their existing disposal methods are more expensive. In these cases, the benefits should be quantified by the amount of such savings.
2. There may be a qualitative improvement in the service provided. In this case the methodology of Contingent Valuation (CVM) may be used to measure the value of those qualitative improvements. In a particular project in a Caribbean country, the prevention of decline and recovery of fishery and the prevention of beach erosion were included as environmental benefits.

Of course, in any one project all of these benefit types might be observed and counted.

While in principle the method of CVM could measure both of the benefits indicated above for wastewater services, it is best to measure the two effects separately if the separation is possible. Separate measurement is preferred because an explicit calculation of cost savings may be more accurate that the implicit value given by respondents in a contingent survey.

The CVM has been in use by IADB since 1988; the first project was the Montevideo Sewerage Master Plan. Recently, it has been applied to Jamaica's Solid Waste Master Plan, financed by the Nordic Fund, and also for the Panama's Sewerage Master Plan, financed by IADB. The CVM is used to obtain information about the WTP for services that have no clear market prices. The CVM has two main characteristics. First, information about the WTP of potential beneficiaries of a project is obtained by presenting to the beneficiary the intended project, explaining its benefits to him, and capturing his WTP in the hypothetical (contingent) case it were built. Second, this information is usually not acquired through the formulation of open questions (asking them directly how much are willing to pay) but under the referendum format, in which each interviewee is shown a hypothetical price to pay if the project is to be built, forcing the potential beneficiary to choose "yes" or "no"). Prices are carefully selected and assigned at random between different beneficiaries. In some cases more than one price may be shown to each respondent.

To explore the plausible range of WTP by the interviewees, a variety of alternative amounts to pay must be presented to different beneficiaries to cover the relevant range. Under the referendum format, since one only obtains a "yes" or "no" response to a given price, the information is limited, therefore the number of interviews required is larger for this format under CVM than that needed by other formats..

Empirical studies have concluded that the fact that the questions are hypothetical does not make the responses arbitrary or biased as long as the questions are formulated carefully. Results obtained show a degree of consistency that indicates that the method is reliable. Also, by presenting the interviewee with alternatives that can be accepted or refused, the answers are much less influenced by strategic considerations. IADB published in 1998, the experience gained by the institution applying this method during the prior 10 years[(34)].

For each proposed alternative there is a possibility that the interviewee will respond affirmatively or negatively. By analyzing all the responses of a group of carefully selected interviewees it is possible to estimate the probability that the benefits to be measured will out weigh the cost they would hypothetically entail. Naturally, the lower the price proposed, the higher the probability that the beneficiary population will accept the corresponding project costs. From all answers to be randomly assigned choices, the median WTP can be calculated.

Special econometric models have been developed to analyze situations in which a group of explanatory variables can be related to the probability of how a consumer will make binary decisions (yes or no). These quantitative models generate probabilities in the continuum between "zero" and "one," though by definition consumer responses can only be one or the other extreme (yes or no). Commonly, these models assume that the probability of responses follows a logistical distribution. In these cases the model is identified as a "logit" model.

The effective WTP by the beneficiaries can be estimated following an approach with the cited statistical techniques that compare situations "without benefits from the project and no additional cost" to the alternative of "with project benefits and additional cost." Abstracting from other variables (ceteris paribus) that may have an effect on the WTP (such as family income levels), it would be expected that the probability of positive responses diminish with the additional cost to the users. The essence of this approach can be explained assuming a linear utility function with a constant marginal utility of income and ignoring other externalities that could influence the decision.

Comparing between the utilities with and without the project allows the analyst to infer that those who give a positive response derive from the existence of the project a higher level of satisfaction than was previously available without the services of the project. That is, the utility of the income foregone is less than the utility gained from the benefits received.

This section presents the steps that must be taken in order to conduct a successful household survey whose purpose is to establish the WTP by users for improved wastewater services using the CVM.

1. Define the sampling frame. In this step, the universe of the population to be surveyed must be established. A land registry of homes or a list of homes with potable water connections in the area where the project is going to be built would provide a suitable frame.
2. Select the sampling strategy. Based on the information available on the sampling frame, a sampling strategy must be designed that would result in the highest possible sampling efficiency, that is, that will produce the least possible variance of results obtained at a given cost.
3. Establish "focal groups." Focal groups must be organized in order to help in the process of designing a CVM questionnaire, to verify the feasibility of applying this approach, and to obtain information of possible prices to be used in the question related to WTP.
4. Develop the preliminary questionnaire. Prepare a pilot questionnaire that will contain all the required questions to establish the WTP for the services.
5. Administration of a pilot survey. A pilot survey will be carried out for the double purpose of testing the questionnaire, and to obtain information that would help to refine the sampling strategy.
6. Design of final questionnaire based on the results of the pilot survey(s) a final questionnaire is designed.

7. Selection of the sample. Following the sampling strategy defined earlier, and with the adjustments suggested by the previous task, the final sample will be selected, making sure to select also alternate potential interviewees, to be used in case that a 100% response ratio is not reached for the original sample.
8. Develop instructions for interviewers. Instructions must be prepared for the interviewers. These will explain all relevant aspects to the interview process (including the process for supervision and verification), and provide answers to any questions that might be raised by the questionnaire. It is important to make sure that the interviewers do not affect the answers by intonation, interpretation, pressure, or invention (sitting on the street corner filling in the surrogate answers).
9. Training for interviewers. The selected interviewers and supervisors will be trained, explaining the purpose of the survey and the methodology for its administration and supervision.
10. Administer the survey. In this step the field work will take place, including the necessary supervision of interviewers and editing of responses to assure their completeness, coherence and quality.
11. Coding and tabulating. Results will be coded, verifying that no mistakes were made during the survey. Ideally, this step should be done at the same time as the results of the survey are coming in, so that in case of errors, the affected interviews can be redone. Subsequently, the codified results will be entered into a computer. The process should be documented to facilitate processing and/or verification.

All benefits should be valued at prices reflecting the opportunity costs of resources to the country. The flow of benefits using the above described model will be disaggregated to a level compatible with the requirements for "optimal" dimensioning of the various components of the project. Geographic and socioeconomic conditions will be considered for this exercise.

In summary two different surveys will be carried out as described above to properly apply the water and wastewater models to measure the benefits of the projects.

Economic or Efficiency Cost Methodology

To obtain the economic cost needed for project evaluation, the following items are considered:

1. investment costs, by year, separated into the following categories: foreign exchange, local equipment (very limited), local materials, skilled and unskilled labor, land and energy.
2. periodic replacement costs by year.
3. incremental costs, by year, covering administration, operation and maintenance.
4. other costs like house connections, land, contingencies, engineering and administration, rights of way, etc Financial costs , inflation, depreciation and taxes will not be included.

All costs have to reflect the opportunity costs of the resources in the particular country. For that purpose, the market costs will be converted to frontier costs using the corresponding conversion factor for tradable and not tradable commodities. This conversion factor is available in IADB's publications.

Long Range Investment Plan

At the prefeasibility level, different technical alternatives satisfying the demand of water and wastewater projects can be set up in order to compare costs and select the minimum cost solution. Traditionally, the minimum cost solution has been limited to comparing investment and maintenance costs, for instance in water projects using: a) different water sources, b) technologies used in designing the treatment facilities, and, c) location of different components of the system. The basis for establishing the different alternatives has also been limited to those projects satisfying the demand during the planning period.

Demand functions provide the tools to estimate the cost of not entirely satisfying demand in one particular point in time. Usage of demand functions in the minimum cost analysis lets the planner choose the "least costly alternative" not only with respect to investment costs but also with regard to the cost to society of only partially satisfying demand at a point in time. A much wider array of alternative investment possibilities thus becomes available in the analysis, since an alternative that by chance does not cover the demand of particular period is not automatically discarded. Besides, it is taken into account with an additional cost assigned as a penalty for each cubic meter of water it is not able to supply. In fact, the demand function properly reflects rationing costs to the user in a medium or long term context in which the consumer is allowed to adjust his consumption patterns to the expected scarcity level.

This system allows other alternatives, such as the loss reduction program (UFW) and staging of large projects to distribute investment costs through time, to be considered, even if demand has to be rationed during a particular point in time. What becomes crucial in the decision of whether to sacrifice water consumption or not is the comparison of failure costs vis-a-vis incremental investment and operational cost of satisfying demand at all times.

Once a minimum cost expansion path is chosen, it should be evaluated to ascertain whether the benefits provided by its implementation are larger or equal to the costs involved. Once again in this evaluation exercise, demand curves are critical in determining benefits, since each point on the curve represents the WTP and therefore the "use value" assigned to that unit of water.

If benefits are shown to be greater than costs, the planner should recommend that particular expansion path at the original tariff level. If benefits of the minimum cost alternative do not override costs pricing levels should be checked against forward looking marginal costs.

If tariffs are lower than marginal costs they should be increased in order to contract future demand levels, and in that manner decrease the cost of supplying it. These adjustments are carried out until the expansion path designed is both a minimum cost solution and is shown to have a positive impact on welfare of society (NPV positive). It is important to note that a positive net benefit does not necessarily mean that the chosen expansion is optimal from a purely economic standpoint, which in addition requires that prices be set to marginal cost. There are, however, valid practical administrative and financial constrains which generally preclude the use of long run marginal cost pricing.

The expansion plan for water and wastewater must comply with the condition that it maximizes the Net Present Value of the proposed investments program. In addition, the following elements must be analyzed: optimal spatial location of the facilities and beneficiaries coverage, optimal selection of technically feasible alternatives, and finally, optimal timing of the investment program considering not only technical restrictions but financial as well.

The Expansion Plan generally covers 20 years and the First Stage approximately 10 years.

First Stage of a Project. The First Stage includes water and wastewater that are considered inseparable and will also include all the technical elements, design, estimate of construction and equipment costs as well as operation and maintenance (O&M) costs. The cost estimate will also include the purchase of water meters, cost of land, house connections, and a technical cooperation program that includes training and public information components.

A prefeasibility study will be carried out for the most attractive combinations of source of supply, transmission, treatment and distribution alternatives for the water subproject, and collection, transmission, pumping and treatment for the wastewater subproject.

Economic Analysis of the First Stage. Once the minimum cost expansion plan is defined by stages and within the proper First Stage the socioeconomic analysis will include profitability analysis performed using the following criteria:

1. NPV calculated using a discount rate of 12% or other rate that properly measures the opportunity cost of capital to the country.
2. Internal Rate of Return (IRR) that must be over 12%.
3. Sensitivity and risk analysis on the NPV and IRR for variations in the main parameters and assumptions utilized.

The benefits will be calculated comparing the proposed First Stage to and without project scenario, as described before.

Water Rates. To design a water rates system that is economically efficient and allows the entity to obtain a reasonable return over its investment, the methodology of Long Run Marginal (Incremental) Cost, should be applied. Always, a compromise between the pure economic efficiency and the return over the investment and the sustainability of the enterprise is made. Financial rates should also be compatible with economic efficiency and the affordability to pay of the users. Commonly people can allow between 3 to 5% of their income to pay for water and sewerage service.

Alternatively, a well designed subsidy system must be in place to assure that low income families are able to meet the water bill.

Ability to Pay and Poverty Targeted Investment. Analysis of the ability to pay of the beneficiaries for the proposed program will be determined. For this objective the following aspects will be considered:

1. the income distribution
2. current and projected tariff level and charges to be applied to consumers

In the event that a large part of the population is not in the position to pay the marginal cost of water, the expansion plan will need to be reevaluated as already mentioned in other part of this document.

For low income analysis for the First Stage, the IADB methodology requires that the impact of the proposed projects on the low income groups be measured. The measurements accepted for this purpose is defined as the ratio of benefits obtained by the low income families, relative to the total benefits accruing accruing to the private sector as a consequence of the project. The benefits of any project can be separated into the benefits perceived by the private and public sectors. From an income distribution point of view it is assumed that the benefits perceived by the public sector are neutral. Thus the question of distribution impact centers on a distribution of benefits in the private sector between the low income people and the rest of the population.

For each country, the IADB defines a threshold income level. Individuals or families with income below the threshold level are considered to be low income. The threshold income level is periodically updated by IADB.

Two types of benefits can be quantified for purposes of estimating benefits to the private sector. On one hand, the direct benefits generated by the project as measured in the economic cost-benefit analysis and, secondly, any distributional effect deriving from discrepancy between market prices and economic opportunity costs, measured at shadow prices. This latter measurement is most germane to low income people employed as laborers in the construction phases of the project. Any discrepancy between their actual wages and the opportunity cost of the labor provided (also known as a shadow price of labor), is to be considered a transfer to the laborer and therefore, should be added to the benefits of the low income groups for the purposes of computing the distributional impact coefficient.

Economic Evaluation for Universal Water Metering and Rehabilitation Programs. The current IADB policy in the sector clearly assigns high priority to implement unaccounted for water (UFW) studies, universal metering and rehabilitation of the existing infrastructure. These components should be considered before the expansion plan is implemented.

The recent IADB publication *"Spilled Water: Institutional Commitment in the Provision of Water Services,"* presents a very simple and effective way to measure the cost-benefit analysis for water losses reduction programs. It is clearly shown that this kind of project have a positive NPV, using a discount rate of 12% and an EIRR over 12%.

Rehabilitation and optimizing of the current infrastructure in water and sewerage project is also stressed by IADB's policy. In IADB terms, rehabilitation means "restore the infrastructure to its original capacity." Before a decision is taken a cost-benefit analysis should be performed following the guidelines already set up in this document. Appendix II includes, as an example, the TOR for an UFW study (Cuenca, Ecuador).

Chapter Four

Water for the Poor

Water for the Poor

General Considerations

Poverty is a secular problem all over the world. At country level, all countries have their own policies for a gradual reduction of poverty. Poverty is not equally distributed among the continents. Some suffer more than others like in Africa, where the World Bank concentrates investments and technical assistance. Poverty, is an endless problem and many efforts have been made at international level to stress the need to alleviate poverty. The Millennium Development Goals establishes the following targets:

1. Between 1990 and 2015, halve the proportion of people whose income is less than US$ 1 a day and the proportion of people who suffer from hunger.
2. Achieve universal primary education by 2015.
3. Eliminate gender disparity in primary and secondary education, preferably by 2005, and in all levels of education no later than 2015.
4. Between 1990 and 2015, reduce the under-3 mortality rate by two-thirds.
5. Between 1990 and 2015, reduce the maternal mortality rate by three-fourths.
6. By 2015, have halted and begun to reverse the spread of HIV/AIDS and have halted and begun to reverse the incidence of malaria and other major diseases.
7. Ensure environmental sustainability by a way of a) integrate the principles of sustainable development into country policies and programmes, and reversing the losses of environmental resources, b) reduce by half at the year 2015, the proportion of people without

sustainable access to drinking water, and c) achieve significant improvement in lives of at least 100 million slum dwellers by 2020.

8. Develop a global partnership for development.

Low income people are located in scattered areas of the cities and in the rural areas as well. The provision of water services for low income people in rural areas has old and known solutions because there is a secular tradition of community participation. Standpipes and pit latrines have been the technology used to provide these basic services. In most of the countries in Latin America rural villages are spread all over the country demanding individual technical solutions tailored to the economic capacity of the users as described in Chapters Three and Five. In some projects, depending on the cost efficiency analysis they are served through extensions of the urban services.

Consequently, this chapter is only devoted to describe the possible solutions to serve the low income persons settled in/near the urban areas. One of the documents consulted among others is a study made by the World Bank [35] that addresses the problem in Sub-Saharan Africa.

BASIC CRITERIA AND ALTERNATIVES

LOW INCOME PEOPLE LOUDLY REQUEST WATER SERVICES. In the majority of the situations in Latin America, there is a misconception that low income persons , cannot afford to pay to get access to the services. As Mr. Hernando de Soto in his famous book, "*The Mystery of Capital*"[36] points out, the low income people in Latin America do have economic potential to support paying for the water services they deserve. Favelas in Río de Janeiro, Brazil, and Pueblos Jovenes in Lima, Peru, among other cities, are good examples. For these persons priority number one is to build a niche where they may live with their families. Several government policies address this essential need and facilitate easy access to get construction materials through formal entities and special banks. Priority number two is to get water. In Peru, there is a formal mechanism by which a citizen association can get credit to build their own water facilities contracting qualified contractors, duly supervise by the bank. This methodology has proven to be satisfactory. Subsidy for low income people should be carefully analyzed case by case and the subsidy has to be transparent. Amid, it is important to note that the cross subsidy among users should be avoided due to the economic inefficiency produced for not applying the marginal cost criteria. Chile is a good example of the use of a transparent methodology. In effect the government identifies first the list of the low income people and then using national or municipal budget, the users receive the money to pay the cost billed by the private company

There are several technical alternatives to provide water and wastewater to the low income people. Among them, three were selected for discussion.

Alternative No 1–Drilling Private Wells. In some communities where there is groundwater available near the surface people drill their own wells, install a pump and store the water over the roof of the houses as an interim solution. However, there is no economy of scale, operation and maintenance is expensive and the quality of the water is not guaranteed. Currently this service is not regulated. There is always room for private initiative to sell pumps and to maintain them. In one particular area of Buenos Aires, this was the initial solution for a new urban development until they realized that they were drinking water with high nitrate content that affects the blood hemoglobin, and can mean the water is contaminated by human feces. The wastewater solution for this scheme is the septic tank that can add more groundwater contamination. The other alternative for wastewater disposal is storage and periodic disposal. In several cities in Uruguay and Argentina this technology is implemented by private companies. The final disposal point is the city's wastewater treatment plant. The charge for the service used is the tipping fee method similar to solid waste disposal.

Alternative No 2–Vendors. When the water utility can not extend the city network to provide water to the non-piped people, is very common that they buy water from "vendors." In some cases the vendors pay for potable water from the installations of the water utility and then transport by truck and resell it charging high prices. The reaction of the people is to reduce their demand to the essential need, for drinking only. In other cases, the vendors build their own, most of the time informal facility, without quality control. A well, a pump a storage tank and trucks are the common solution when groundwater is available. People realize that this form of service is the only solution to provide water when the location of the settlement is very far from the municipal water network. However, they want to get public service because it is more reliable in quantity and quality and cheaper. In Asunción,

Paraguay water vendors solve the problem for hundreds of people located outside the water network.

The water vendors, encouraged to develop as an elemental form of privatization, are an incomplete solution because there is no guarantee of water quality and there is no regulation. For wastewater disposal the septic tank technology is very common but requires maintenance that may be offered by private companies. This solution can not be applied when the groundwater level is too near the surface.

Alternative 3–Simplified Systems. It is always technically possible to design simplified water and wastewater systems at low cost. A water service delivered by standpipes is a technical solution and it is the first step from where a more elaborate system can be developed based on "community will." The public standpipe system can be equipped with water meters and the water can be sold based on consumption.

A wastewater collection system based on condominial or small bore systems may be a perfect technical solution (see Chapter Three). The community is responsible not only for the design and construction of the system but to maintain and keep it running. The cost of the wastewater unit is about 50% of the cost of a conventional system. In both solutions there is always room for the private initiative. In the condominial or small bore wastewater systems, maintenance can be privatized as well as the operation of a simple wastewater treatment plant.

Policies and Strategies

To better serve the low income people it is necessary to develop new policies and strategies. The government, state and municipal entities should be embodied in the implementation of this task. The main requirements are discussed below.

Legal Framework. Reform of the water law is indispensable to incorporate innovative cost effective solutions. It should be mandatory that the public or private water utility is responsible to incorporate these systems within its service area. Furthermore, the regulating agency should extend its mandate to control the quality of the services and to penalize the infractions. Additional expenses incurred by the regulatory agency to comply with this new commitment should be considered in its general budget. Designs and construction of the works should follow the public bid procedure and contracts with private companies should be supervised and receive the approval of the water utility.

Institutional Framework. Regardless of the alternative selected there always must be a formal institution or a group of community representatives to bolster the implementation of the service. The managing team should be democratically elected and should receive from the water utility basic guidance to:

1. understand the characteristics of the model adopted,
2. acknowledge the importance of preserving water quality and the environment,
3. apply the best maintenance practices to sustain the service, and,
4. accept the economic commitment required to get clean water.

Technical Standards. The water utility should incorporate into its standards the necessary changes to facilitate the design of innovative solutions. New design parameters should be developed and tested that are aimed at preserving water quality and the environment.

Community Education. To facilitate the implementation of the model it is necessary that the water utility develop special programs to educate the users about their rights and duties, quality standards, protection of the environment and why they have to pay to receive services.

Manpower. There must be a program supported by the water utility to develop manpower capability. Eventually, some of the trained people of the community might become private operators.

Financing of Capital. There are several sources to finance the construction of the necessary works to serve the urban poor like: grants and loans, taxes and revenues, NGO, community and users contributions and private funds. Generally, the users do not finance the major capital costs: production, treatment, storage, and primary network. Proper financing of capital should be resolved case by case depending upon the sources available, the cost of the project and the practices in the country.

Chapter Five

Current Institutional Models in Latin America

Current Institutional Models in Latin America

Introduction

The information, findings, conclusions set forth below are based on four (4) cases:

Case 1. Survey of the situation of privatization of water and wastewater utilities for 12 selected countries[37]

Case 2. The administration of water supply and sanitation in five enterprises in three countries[46]

Case 3. The water business in Latin America. A new approach?[38]

Case 4. The role of private sector in providing water to developing countries [39]

Case 1. Survey of the situation of privatization of water and wastewater utilities for 12 selected countries

Argentina

Situation in the sector. Only 69% of the urban population has connections for water at home, while only 17% of the rural population has any form of sewerage. There are also regional disparities; the province of Cordova has 84% coverage for water and 88% for sewerage, while for Misiones the figures are 43% and 50%, respectively. Moreover, only 27% of the sewage that enters the system receives primary treatment, and only 8% secondary treatment. The water losses in the urban systems are over 50%. There is underway for many years a national program very successful and supported by the IADB for rural areas . The main characteristic is the creation of Co-Operatives (Juntas) to manage the systems. The user has to contribute 50% of the total cost of the projects. The other 50% comes from the province. Furthermore, the Juntas operate the systems and establish the water rates.

Decentralization of services. In 1980, the government decentralized control of the water and sanitation that was in the Obras Sanitarias de la Nacíon (OSN), delegating it to the provinces. At first this was of little help as the provincial systems were always government monopolies, although some had a degree of operational autonomy. However; in the provinces of El Chaco, Chubut, La Pampa, and San Luis, the majority of the services are run by Co-Operatives, while a few smaller municipalities have always operated their own services. Currently, there are approximately 1,500 providers of water and sewage services in Argentina. Currently ENOHSA is the entity responsible for the urban and rural water programs.

Privatization for urban systems. In 1989, the government embarked on a major privatization program in the country and water and sewage were not excluded. Buenos Aires, Corrientes, Formosa, Santa Fe, and Tucuman transferred their services to private concessions, while Cordova planed to follow the trend. In Tucuman, the concession awarded to a French operator was canceled for political reasons. Nowadays a management contract with a Spanish operator is in effect.

In 1993. Buenos Aires contracted a 30 year concession with a French operator Lyonnaise des Eaux creating a new enterprise, Aguas Argentinas. The new enterprise offered an ambitious program of US$ 150 millions. An IADB loan for US$ 98 millions to expand the water services in the western area of the city was redesigned to contribute the international funds for the concession. Water meters were installed on only 20% of the connections and few of these worked properly. Therefore, water users had no incentive to conserve water. Consumption was estimated at approximately 400-500 liters per person per day, two or three times as high as normal in a well metered system. The water losses were as high as 40%. The ratio of number of employees per 1000 house connections was extremely high in the order of 10.

Aguas Argentinas offered to increase in a 30 year period, the water supply coverage to 100% and sewerage coverage to 90%, a goal to which it has already made some progress. The company indicated that it has increased the capacity of its water treatment plant (Palermo plant) by 30%. The only wastewater treatment plant, located in the south was operated successfully after simple modifications. By means of a few thousand early retirement packages, the number of employees per 1,000 connections was reduced to 3.5.

Nevertheless, a new controversy regarding the company has recently arisen. Aguas Argentinas found, in 1997, that the stipulation requiring it to make new connections was ruining its profitability. The government, in order to continue with the promised new connections, decided to allow prices to increase. Naturally this did not go down well with consumers who protested loudly. An independent regulatory authority, that is crucial for the success of any privatization, has been set up. Many problems and potential problems plague the agency, known by the acronym ETOSS. Its commission must represent three different levels of government: the national government, which owns the assets, the province, and the municipal government of Buenos Aires. This entity was created after the concession contract was signed and resulted in a continued quarrel with the operator. Besides, to get the agreement of 3 different powers made its tasks rather difficult.

Bolivia

Situation in the sector. The water and sewerage situation in Bolivia for urban centers is marked by a disparity between urban and rural areas. In urban areas, 84% of the residents are connected to water pipes, and 63% to sewerage, as opposed to 24% and 17%, respectively, in rural areas. Overall, only 58% of the people are connected to water supply, and 43% to sewers, some of the lowest levels in South America.

Decentralization of services. In 1991, Bolivia began a major restructuring of the sector, which involved the transfer of powers from the central level to the municipal level. As expected, this has led to much experimentation. For example, just recently, the IADB approved a US$ 70 million loan to reform regulatory systems so as to encourage private sector involvement.

Thus far, private participation remains incipient, hindered by jurisdictional conflicts. These have often occurred because the central government has tried to dictate terms to the towns, some of whom have, naturally, not responded in a positive way. Nevertheless La Paz, the capital, and El Alto turned their water and sewerage systems over to the French company Lyonnaise des Eaux in July 1997, despite large protests and agitations by the opposition, which periodically paralyzed both municipalities. Lyonnaise des Eaux owns 34% of the new company, while a combination of Bolivian and Argentine directors own the rest. This concesion has been recently canceled due to social protests.

The rural areas of the Altiplano region have been the subject of a World Bank project called Yacupaj, meaning "for the water" in Quechua. First conceived in 1990, its supporters claim it represents a new type of development that listens more closely to the common people and therefore caters more to their desires and needs. The results of Yacupaj will pave the way for a much larger investment program, called PROSABAR, which is expected to benefit 800,000 people. Yacupaj, serving modest communities with 50 to 250 inhabitants, provided rudimentary hand pumps and latrines. Most of the workers on the project were locals who were specially trained to do the work. According to the project's promoters, all the communities affected were carefully consulted in advance.

After the project was completed, the results were analyzed in order to discern the preferences of the villagers who will form the basis of PROSABAR. This investment program is only in its early stages. It is supposed to be using the same techniques, but on a larger scale.

Brazil

Situation in the sector. In Brazil, 71% of the population, nationwide, are served by the water supply system, but only 35% are connected to public sewers with 49% having only rudimentary cesspools or no means of disposal at all. 80% of the sewage throughout the country does not receive any treatment. For rural areas a simplified water system has been built. For sewage the "in situ" disposal method (on site) is adopted.

Privatization. Recently, some of the municipal governments that own the water and sewage systems have sought private sector help. The trend is most prevalent in São Paulo, wealthiest state in per capita income. So far, the largest concession has been granted to Aguas de Limeira, a joint venture between the French company Lyonnaise des Eaux and Companhia Brasileira de Projectos e Obras, a Brazilian civil construction firm, providing water and sanitation to the 250,000 people of the São Paulo suburb of Limeira. Ribeirao Preto. Another important city in São Paulo awarded the concession to CH2M and REK for the design, construction and operation of the municipality's wastewater treatment system. Río de Janeiro, is reportedly considering a concession as well. In 1998, Suez-Lyonnaise des Eaux joined up with Vivendi, a Lyonnaise subsidiary, and Degremont, to build two water treatment plants in São Paulo: one for São Miguel (population 700,000) and one for Novo Mondo (population 1,000,000). Because of its large size and inadequate infrastructure, Brazil offers many opportunities for private sector participation.

Chile

Situation of the sector. The water coverage in Chile is excellent by Latin American standards. In urban areas, where most of the people live, 99% of the households are connected to drinking water, and 89% to sewerage. In the more densely populated rural areas, 82% have water supply, but in the more dispersed country settlements, only 14% do. In urban systems water losses average 31%. Right now, only 14% of the sewage is treated, but the government plans to have universal treatment of all sewage that enters the system by 2005. The experience indicates that this target is very difficult to achieve. For rural areas cooperatives have been created operating very satisfactorily following the first loan granted by IADB in 1974.

Privatization. The most interesting aspect of water policy in Chile is that it is the only country in the region to have created a market in water rights, due to a policy change made in 1981. In other countries, the government owns the water and may sell the rights to its use and distribution to a private party by contract. However, the terms of the contract are always such that the private company can either finish the contract, after which the resource returns to government ownership, or sell the unexploited rights directly back to the government before the contract expires. In Chile, water rights are owned by private citizens or businesses, who may lease or sell the rights to other private parties. In theory, this system is supposed to encourage efficient use and conservation by making water a valuable resource for the party that owns it. The government's role is limited to granting concessions, financing certain parts of the infrastructure, and enforcing property rights.

Since 1989, Chile's municipal water companies have been run by concessions. There are 13 different regional companies for the 12 political regions and Santiago, each granting four separate concessions: i) for water supply and treatment, ii) for water distribution, iii) for operation of sewers, and, iv) for sewage treatment. All operators are kept under close scrutiny by the Superintendencia de Servicios Sanitarios (SSS), a strong and autonomous government agency.

In December 1996, the government introduced a bill to fully privatize state-run water works, but it faced strong opposition. After much political debate the bill was finally passed, albeit with some compromises, including a stipulation that the government must maintain 35% equity, with some of the remainder being owned by the company employees. In April 1997, the government announced its intention to privatize wastewater treatment as well. The privatization law was finally approved in January 1998.

Colombia

Situation in the sector. The level of water supply in Colombia is quite high, at approximately 90% for both urban and rural areas. But a large portion of the water supplied is not potable. Only 62% of the urban and 10% of the rural population have "access to water" that has been treated for human consumption. While 70% of the urban population has acceptable sanitation, only 27% of the rural population does. The water systems suffer from excessively high water losses ranging from 30% to 50%. A negligible 2% of the sewage receives treatment.

Privatization. In 1992, in keeping with a general privatization program, the government introduced law 142/94, intended to establish a proper regulatory framework for private participation. Private sector involvement, non-existent before this law, is small but is expected to expand.

The role of the private sector in water and sewage is already evident in large Colombian cities. The public utility in Barranquilla is run by a mixed corporation with a 13% private stake and even makes a profit. In Cartagena, the newly privatized company dismissed half the workforce, presumably because the utility was overstaffed.

The World Bank, estimating that only half of all the systems nationwide properly treat water for human consumption, and that even fewer do so consistently, has encouraged private participation to remedy these problems. It has offered the biggest loans to Bogotá, the capital city, which accepted two loans for $145 million in 1995 as part of a $717 million project to convert the municipal monopoly, Bogotá water supply and sewerage, into a profitable, commercially run utility, and to reduce the pollution of the wetlands in the Bogotá river basin, which has allegedly been the cause of several wildlife extinctions. As expected, the private sector soon got involved, with Union

Fenosa - ACEX winning a contract two years later with the largest Bogotá utility, serving over a million people, in order to modernize the whole system of management.

Ecuador

Situation in the sector. In Ecuador's urban areas, 78% of the population has water supply and 60% has connections to sewers. In rural areas, however, the corresponding figures are 74% and 8%. The water systems are generally of very low quality; water losses are between 40 and 60% while few deliver either a continuous or a high quality supply.

Municipalization. In cities and towns, the service is provided by the municipalities, no longer by Instituto Ecuatoriano de Obras Sanitarias (IEOS), who usually establish autonomous public companies. In rural areas, approximately 2,000 cooperatives, committees of locals, do the job. The IADB has intervened to encourage further private participation in the sector, giving a grant to the government in order to set up the necessary reforms of pricing and regulatory procedures. The government intends to eventually privatize all water utilities, for the sake of financing further investment.

Privatization. Reforms in the country's two largest cities, Quito and Guayaquil, are already well underway. In Quito, the IADB financed $136 million of a $170 million project in September 1994 to repair and modernize the water and sewage system.

While the loan to Quito is meant to prepare for privatization by improving the physical workings of the system, in Guayaquil the emphasis is more on improving the institutional situation. A smaller US $40 million loan is intended to help the Guayaquil Water and Sewerage Company award the contract. The city certainly needs some major improvements, as, aside from the decrepit condition of the existing infrastructure, coverage for water and sewers respectively is only 63% and 52%

Mexico

Situation in the sector. Overall, 84% of the population receive piped water supply, and 67% are connected to sewers. However, in rural areas, the corresponding figures are 53% and 21%. In addition, the regional inequalities are severe, with only 69% being connected to water and 47% to sewerage in the Southeast, as opposed to 97% and 95%, respectively, in the federal district.

For decades, the official doctrine in Mexico was that water was a basic right that should be provided by the government cheaply, if not free of charge. This policy was not abandoned until 1992, when a new law both encouraged private participation in municipal utilities and assigned water rights for urban water usage. Nowadays some concessions are being offered as well, even in the hinterland; Azurix Corp., a division of Enron, recently acquired a stake in a long term concession offered in Cancun.

For decades, Mexico City did nothing to encourage water conservation. In the mid-1990s, water leakage was between 30 and 40%; unacceptable losses in a city with such a severe water shortage. Although around half of the homes had water meters, many of them were dysfunctional or were not read on a regular basis. Installing meters for the rest of the population would incur costs running into the hundreds of millions of dollars. A 1992 World Bank study found that providing water in Mexico City was very expensive, at $1.00 per cubic meter, but the government of the federal district of Mexico City collected only 10 cents in bills, a tenth of the total cost. The amount collected in Mexico City is significantly less than the figures for many other Mexican cities, such as Monterrey, which collects 37 cents.

Sector reform. Reforms are desperately needed, but political factors make them difficult to implement. During the years of one-party rule, the government ran and subsidized all the utilities as monopolies and propagated the view that water was a right and people could use it as they pleased. Thus, much of the population has now been inculcated with both a belief in, and a habit of, uncontrolled water use. Now that the government has changed the Comision Nacional de Agua is telling its citizens to conserve water. Political opposition to the necessary changes is likely to be intense.

Privatization. By now, all water professionals agree that the only choice is to take measures to increase efficiency and reduce demand, a view set forth in the report entitled "Mexico City's Water Supply." Finally, it appears that the government is taking steps in that direction. In October 1992, in keeping with the new national law being introduced at that time, the federal district of the city asked for bids to run and maintain the city water supply. The city was divided into four zones, each one covering about 250,000 house connections. Management contracts were awarded to 4 different operators from France, England and Spain. These contracts are only for management of the water distribution network and are paid based on efficiency. It is envisioned to award one concession for the whole water system in the future.

In the same year, the new federal water law was passed, which was partly meant to alleviate the water crisis in Mexico City. Before then, anyone was permitted to drill a well in the basin and use the groundwater. However, the law stipulated that anyone wanting to drill a new well would have to buy the rights from an existing user. If properly enforced, and the costs of water rights properly calculated, this should restrict any new wells in heavily exploited areas.

Although the privatization program took a long time to materialize, by 1998 it was finally implemented. The French company Lyonnaise des Eaux is now operating the water and sewerage treatment plants in Southeast Mexico City in a joint venture with a local company, serving 2.5 million people, while the other major French firm Générale des Eaux supplies water to 2.2 million in another part of the city under its Mexican subsidiary, Sapsa.

Paraguay

Situation of the sector. The Corporación de Obras Sanitarias (CORPOSANA), owned by the Paraguayan government, is the sole provider of water and sewerage services in all towns and cities with populations of 4,000 people or more, and it does so very poorly. Only 42% of the population in these towns is connected to water, and 35% to sewers. In Asunción, the figures are much higher, at 86% for water and 68% for sewerage. Only 17% of the rural population have drinking water supply, but 60% have sanitation.

Reform of the sector. Because of its embarrassingly bad record, CORPOSANA is forced to allow private water vendors to serve the large areas of the cities that it does not cover, just to make the situation bearable. These vendors buy water from CORPOSANA and make contracts with a specific group of consumers and even go so far as to install metered house connections. This is an incipient franchise method that it will be discussed in the next chapter. CORPOSANA seems to have concluded that the best way to expand coverage is to privatize many of its functions.

Water for the rural areas. The government is also trying to improve the water supply in rural areas, under jurisdiction of SENASA with the help of a US $40 million World Bank loan. Given the magnitude of the problems, the program seems to serve only 330 communities with a total of 340,000 inhabitants, increasing the number of rural homes with connections from 20% in 1997 to 30% by 2003. Only about 140,000 residents are expected to benefit from better sanitation systems (in situ disposal).

The World Bank's program required that the government contributes only 40% of the investment. The users pay 15% of the capital cost of the sanitation facilities at the time of implementation and contribute further cash, labor, or materials over 10 years, together amounting to an additional 45% of the total capital costs. In addition to that, the users pay the entire operational cost. This is a very demanding investment program for the communities.

Peru

Situation of the sector. The Peruvian water and sanitation situation is poor. Just 71% of the urban population and 31% of the rural population have water supply, while the corresponding figures for sewer connections are 56% and 23%. Those without service often have to pay $7 to $8 a month for four cubic meters from water trucks. The capital city of Lima dumps 17 to 18 cubic meters per second of untreated wastewater into the sea adversely affecting the marine environment.

The quality of the systems is deficient. In the smaller urban centers, only 70% of those served have 10 hours of daily supply, and just 20% have more than 20 hours. The water pressure in 70% of the districts does not meet minimum recommended standards. Overall, fewer than 60% have safe water, and 45% safe sanitation.

Decentralization. Since 1990, the system has been decentralized and private participation encouraged as part of the government's large privatization plan. The Lima system, SEDAPAL, serving five million people in a metropolis of seven million, was supposed to be transferred to concession. A standard 30-year contract was expected, with investment demands being $600 million in the first five years and $1 billion in the first ten years. Even though the privatization process was underway for about one year the president of Peru suddenly cancelled the privatization plan. Instead SEDAPAL received US $400 million investment program partially financed by the World Bank.

The World Bank has also been trying to protect the coastline, now polluted, supporting a US $300 million project, which includes $230 million for environmental infrastructure, $50 million for improved coastal management, and $20 million for a revamped regulatory authority. The government of Peru is carrying a 33% share of the cost, some of which would have been paid for by the winner of the concession, had it taken place.

Municipalization. Outside of the capital, the private sector is also being encouraged to step in. Several laws passed since 1990 have effectively ended the policy of public service monopolies in water and sewerage. The individual municipalities are now in a position to grant concessions to both domestic and foreign firms, and many are expected to do so. Recently, a project to increase the water supply uses a new source of water, the Chillon River. It was awarded by SEDAPAL to a Spaniard consortium .by the way of a BOT arrangement. Peru approved in 1994 a water law to allow private sector participation. A capable regulatory entity was also created and implemented at the same time.

Privatization. The government has created PROINVERSION as an entity in charge of promoting the private sector participation in the sector using either concessions or management contracts with qualified operators.

Trinidad and Tobago.

Situation in the sector. Water supply coverage in Trinidad and Tobago is rather high, with 92% having easy access to piped water at home, on their property, or at communal faucets. However, only 30% are connected to sewers, and almost all of them are in urban areas. Much of the urban and almost the entire rural population rely on septic tanks or pit latrines.

Trinidad receives about 200 centimeters of rain annually, which has allowed it to afford the dreadful inefficiency of its water supply system, which can account for only half of its water. The other half is unaccounted for water that disappears due to leaky pipes, illegal connections, illegitimate use of hydrants, waste at communal faucets, and unregistered users. In the early 1990s, only 80% of the connected properties actually paid their bills. Only 26% of the customers received water 24 hours a day, and only 62% received water 12 hours a day. The water supplied was not very clean, making water-related diseases a leading cause of death in the country. The utility, Water and Sewerage Authority (WASA) was not able to set its own reasonable rates; the public utilities commission only granted it two price increases in 32 years. There were 12 staff per 1,000 connections, a very high number indicating the degree of inefficiency and bureaucracy.

Privatization. Given the sorry state of its system, it was not surprising that the government chose to partially privatize WASA in 1995. The winner of the short-term, three-year contract was Severn Trent Water International, a subsidiary of Severn Trent in the U.K., which owns the new Trinidad and Tobago Water Services along with WASA. It had only until 1998 to prove itself by meeting specific goals, such as achieving financial viability without reducing staff or raising prices.

The World Bank has been closely involved with the privatization process. It has helped the government transform WASA into a commercially oriented company and organize bids to repair the decrepit infrastructure. It is also offering a US $65 million loan for a $90 million project to detect leaks and monitor pressure inside the system, install metering, and improve the existing sewage treatment plants.

Uruguay

Situation in the sector. Access to safe water in Uruguay is high, about 88%. Although 92% have sanitation, only 48% are linked to the sewers, and the on-site sanitation is normally considered inadequate. There are also inequalities in the quantity and quality of coverage, which is better in the capital, Montevideo, than in the provincial towns or rural areas. Many residents have to suffer restrictions on supply during the summer months.

Privatization. Uruguay is taking loans from the World Bank to improve its urban systems, but has not yet committed itself to granting concessions. The national entity Obras Sanitarias del Estado has passed an extended assistance program. Some small privately owned companies run systems in the country. In 1995, a national referendum refused any form of privatization.

Venezuela

Situation in the sector. In urban areas, where over 90% of the Venezuelan population live, 68% of the residents have safe drinking water, 51% through connections at home, and 55% have sanitation services, 33% through the sewers. In rural areas, 67% have water, and 59% sanitation. The quality of most services is poor and the sewage does not receive proper treatment.

The water situation in the capital, Caracas, is serious. The population of the city has rapidly increased to over five million without a corresponding increase in the water supply system. It is under study for some years to bring water from a source located 200 km from the city. The supply deficit is currently 260,000 cubic meters per day, and a plan formulated in 1996 to draw water from 50 wells only promised to reduce this figure by 43%. The problems are made worse by mismanagement and massive water losses from the distribution system.

Privatization. A fiasco occurred in 1992 when the government offered a 25-year concession for the water service in Caracas, but not one bid was submitted by the five contenders before the deadline. The government was in such a hurry to offer the concession that it did so before a proper framework was in place. There were 2 main problems, the water rate of US$ 0.04 per cubic meter was considered far too low, and the lack of diligence from the regulatory commission, (the MANCOMUNIDAD), an association of municipalities made up of municipal representatives who disliked one another.

Outside of Caracas, efforts are being made to improve the water supply and sanitation systems, which suffer from rationing, low pressure, poor quality, absence of sewage treatment facilities, and financial mismanagement, first by a national monopoly HIDROVEN and then by the regional monopolies created in 1990. Private participation remains very limited. The World Bank began in November 1995 a US $70 million project in the Monagas state, which is expected to rehabilitate the existing infrastructure, to improve wastewater treatment, and to transform the local water and sewerage company, HIDROCARIBE into a commercially run utility.

Summary of the Case 1 Countries' Strategies

A summary of the models adopted for urban areas in the selected cities is presented in the following Table.

Country	City/State	Urban Model	Comments
Argentina	Buenos Aires	Full privatization awarded to a French company	Continuos quarrel between the concessionaire and the regulator.
	Corrientes	Full privatization	
	Formosa	Full privatization	
	Santa Fe	Full privatization	
	Tucuman	Management contract with a Spaniard Operator	Full privatization awarded to a French company was canceled.
	Cordova	Full privatization	In process

Water for All in Latin America: *A Daunting Task*

Country	City/State	Urban Model	Comments
Bolivia	La Paz	Privatization for El Alto, a small suburb of La Paz. Awarded to a French company that owns 34%. The rest belongs to 2 companies from Bolivia and Argentina. Recently, this concesion was canceled.	Trend to full privatization of all the water utilities at country level.
Brazil	São Paulo	Privatization for Limeira, a suburb of São Paulo, awarded to a French company.	Trend to full privatization of all the water utilities at country level.
	Ribeirao Pretto	Privatization for this São Paulo's important area awarded to an USA company associated to a local company	
	Río de Janeiro	Privatization to build and operate 2 water treatment plants for 2 areas of the city, San Miguel and Novo Mondo, awarded to a French company	
Chile	Santiago and 12 political regions	Full privatization	Process is underway. .Superintendencia de Servicios Sanitarios (SSS) is a very strong and capable Regulator.
Colombia	Barranquilla	Privatization awarded to a mixed corporation with 13% private stake	Trend to full privatization at country level.
	Cartagena	Full privatization	
	Bogotá	Management contract to operate the water system awarded to a Spaniard company in association with a local company	
Ecuador	Quito	Loan from IADB to improve the services. Full privatization will come later	Trend to Privatization and management contracts at country level.
	Guayaquil	Loan from IADB to institutional capacity of the local company	
Mexico	Mexico City	Water distribution system was divided in 4 areas of 250,000 house connections and awarded 4 management contracts to French, English and Spain companies	Trend to full privatization .at country level.
Paraguay	Asunción	Water vendors in parts of the city	Trend to privatization and or management contracts at country level.

Country	City/State	Urban Model	Comments
Peru	Lima	Full privatization for Lima's utility was cancelled. The World Bank is financing a large contract awarded to Sedapal, the local utility, to improve the water and sewerage systems. Privatization of the Chillon's river new source of water supply for Lima was awarded to a Spanish corporation.	Trend to full privatization at country level, except for Lima.
Trinidad and Tobago	Port Spain	Mixed corporation with an English company. A three year contract was awarded to operate and improve the water and sewerage systems	Trend to management contracts at country level.
Uruguay	Montevideo	Institutional Development of Obras Sanitarias del Estado, the water utility at country level	Privatization was rejected in a national Referendum.
	Rincón de la Bolsa	A privatization contract was awarded to a local corporation to manage and operate the water and sewerage systems of this small city located near to Punta del Este, an important turistic resort area.	
Venezuela	Caracas	Privatization of the water and sewerage systems was canceled.	Trend to management contracts at country level.
	Monagas	Contract awarded to improve the capability of the local entity called Hidrocaribe and to develop a construction program.	Trend for privatization of the water and sewerage utility.

Case 2. The administration of water supply and sanitation in 5 enterprises in three countries

In 1996, the IADB's Evaluation Office selected five companies from three countries: Chile, Ecuador and Uruguay and conducted an evaluation of the performance of these institutions. The principal conclusions were:

Strategy Plan. The existence and implementation of a strategy plan is a necessary component for organizational success.

Regulatory Framework. The existence of a clear and stable regulatory framework is indispensable for the companies to work effectively.

Tariff/Rates. The development and implementation of comprehensive and realistic tariff policy is basic for the companies to survive.

Management. The stability and effectiveness of the companies depend upon the quality and continuity of the management.

Human Resources. For the most part, the five companies have no comprehensive human resources management system.

Private Sector Participation. The analysis of the 5 companies provides mixed evidence of the appropriateness of privatization as an effective policy instrument.

CASE 3. WATER BUSINESS IN LATIN AMERICA: A NEW APPROACH?[(38)]

Since Latin America has had little experience in private sector participation for the water supply and wastewater management industry, the leading companies from Europe (particularly England, France and Spain) realized that the water business in Latin America might be profitable and participated in public bids promoted mostly by the World bank and recently by the IADB.

Models adopted for PSP were imported and copied with varied success. For example, for the past five years in Buenos Aires, Argentina, a French group has been conducting a 30-year concession contract with relative success because the government does not have the experience to properly handle the approach or the public's confidence in them to handle it. However, in Tucuman, a province of Argentina, a concession contract did not work due to political intervention and poor tender documents. As a result, the concession was cancelled after just one year of operation.

In Lima, Peru, after almost two years of project preparation, a 30-year concession was put on hold for political reason. In addition, another problem was the impatient government's privatization commission who wanted to finish the process in a single year without political will, public awareness and acceptance.

In Panama, the public participation process has been underway in stages. The first stage ended when the reform law was issued to allow PSP. In the second stage the regulatory body and regulatory framework were completed. Originally a concessionaire approach was considered for two new entities, one for Panama City and the other one for the principal cities of the country. This scheme left the small cities to be managed by municipal enterprises yet to be created. Later on the government decided to prepare an international bid to select a concessionaire for the whole sector, urban and rural. Once the bid documents were ready the government decided to put on hold the process due to political reasons.

In Uruguay, the national referendum on PSP had a negative response. Consequently, the sector reform is on hold until the approval of a new law.

In a November 1977, in a seminar sponsored by the IADB, thirteen cases of PSP in twelve countries were presented. The following Table summarizes the individual country's approaches and length of the processes. The information provided is updated.

Latin American Scorecard for PSP

Country	City/State	PSP Approach	Group	Duration	Comments
Argentina	Buenos Aires	Concession	French	4 years	Too rigid contract that impedes the necessary interchange of views with the concessionaire.
Argentina	Tucuman	Concession	French	1 year	Revoked
Peru	Lima	Concession		3 years	On hold
Colombia	Cartagena	Public+PSP	Spain	2 years	Public owned company, management contract with an international operator
Venezuela	Lara	Public	Local	1 year	Publicly owned company: state and municipalities
Ecuador	Guayaquil	Concession			In process
Panama	Sector	Concession			On hold
El Salvador	Sector	Public	Local	3 years	Central institution and local PSP using management contracts
Bolivia	La Paz	Concession	French	2 years	Only for a small area of the city. Recently canceled.
Mexico	Mexico	Public	French, English, Spanish	4 years	Four management contracts with international operators by stages, toward a single concession in the future
Brazil	São Paulo	Concession	French and USA	2 years	Only for 2 cities Ribeirao Preto and Limeira
Chile	Sector	Concession	French, England, Spain	Several years	By law all the water and sewerage systems must be privatized. Capital mixed, PSP and region.

Country	City/State	PSP Approach	Group	Duration	Comments
Venezuela	Monagas	Public	Spain	1 year	Publicly owned company and a management contract

As shown in the Table, there is not a unique model adopted in Latin America for PSP. Based on experience the following preliminary conclusions can be drawn:

1. Sector reform is a political process. Without political commitment nothing can be done.
2. Public awareness is critical to success. Public resistance to the change should be considered and several alternatives should be explored before a PSP decision is taken. This will help resolve worker complaints about the whole process. In Latin America water has been considered a commodity to be provided by the government free of charge to the people. Mexico City is a good example of the negative public response to the increase in water rates. The rejection of the entire process in Uruguay also is a good example of negative public perception and fear.
3. The role of the international lending agencies should be considered within its proper context. Banks can be used as advisors, but not as promoters of PSP per se. For example, the problems in Lima resulted from too much external intervention from the World Bank, without the political commitment.
4. The PSP approach is used for two main reasons: to reduce the burden on the government's budget and to increase efficiency. However this approach is difficult when unaccounted for water is over 50%, metering is about 25%, billing is only to 40% of the users and billing collection is only 30%. Therefore, the transfer of public obligations to a PSP should be a gradual process. The cities of Lara in Venezuela and Cartagena in Colombia are good examples of this approach.
5. In order for PSP to be attractive to the private sector, there are elements that must already be in place in the country. Some of these elements include sector reform, separation of functions among planning, regulation and operation, the appointment of a sound and powerful regulatory body, and the design of a regulatory framework to define PSP rights and obligations giving special attention to the users' rights and the environment.

6. For any proposed PSP alternative, a sound technical document should be prepared giving priority to rehabilitation. The investment program should be the least cost alternative and take into account the local capability to pay the rates calculated based on long run marginal cost..
7. The Build-Operate-Transfer (BOT) methodology might be a good alternative to attract PSP for specific system components. Panama awarded this king of contract to an international consortium to build a new water treatment plant based on current tariffs that were considered attractive.
8. Awareness of PSP experience in the region is important in order to draw conclusions about the best method. Therefore, the selection of an experienced consultant to accompany the whole process is critical.
9. Tender documents should be clear but not too rigid; periodic reviews should be considered. A 30 year horizon presents too many uncertainties to define *a priori* all aspects of the concession. The regulatory body must be aware of its new role negotiating but not imposing rigid conditions to the concessionaire.
10. A directory of experienced consultants and international operators should be available in the region. Conversations with private operators at early stage of the process do not preclude the calling for an international competition later. Their opinion and experience is valuable and free of charge.

Case 4. The Role of the Private Sector in Providing Water to Developing Countries

Background

Mr. Gabriel Roth[39] wrote in 1985, a very interesting paper about public sector participation in the water sector. He reviewed the experiences in France, India, Dominican Republic, Chile, Bolivia, Argentina, Pakistan and the Philippines among others. Some conclusions of that paper are presented below:

1. The brief review indicates that water supply, especially for drinking, is in a deplorable state in many countries.

2. Because of the universal tendency to politicize and municipalize piped water and sewerage systems, the private sector is unlikely to finance infrastructure in the sector. French affermage is a model that provides management that can be widely copied and developed.
3. In the provision of non-piped systems the private sector can provide vended water in towns and villages all over the world.
4. The situation where rich people receive subsidized piped services, while the poor pay for expensive water from vendors, is likely to remain until governments allow economic markets to work in the water sector.
5. In rural areas, or small cities, local associations or cooperatives can be formed to provide services for which people can pay. In some cases the cheaper alternative is water vending.

Institutional Models–Pros and Cons–A Brief Summary

Central Government owned utilities. This scheme became obsolete. The current trend is to create autonomous entities like Private or State/ Private owned entities. The "pro" of this scheme is its compromise to deliver the services to all. However, the "con" is its acceptance of the cross subsidy that creates inefficiency in the model. Political influence is a common "con".

Municipally owned utilities. This scheme is efficient provided that a municipal owned private entity is created whether is municipal owned or a mix of capital with public and private participation. The "pro", is the active participation of the users in the decision- making process. The "con", is the political control exerted by the municipality because of its economic participation.

Full concessions. This model has been successfully developed in a vast number of cases. Its "pro", is the reduction of the economic burden on the government. The "con", is that this model requires a perfect coordination with the Regulatory entity. Besides, there is a negative public response due to the elimination of personnel to make the entity more efficient. Consequently, once the political decision is taken a intense public awareness campaign is needed.

BOT or BOOT Models. For specific elements of the system like the treatment plants is customarily, in developed countries and in some countries in Latin America, to adopt this model. The "pro" is that the process to contract the private operator is shorten.

Elements of Reform Required

It is proven that the development of the sector to make it sustainable requires drastic changes, as described before, in regard to the institutional model, public participation and financing of the necessary capital. The principal elements of reform of the water sector to allow PSP include ample competence among several actors. The approval of a water law is often essential. In most countries the law proposed by the executive branch requires congress' approval, may require modifications and delays may result.

Approval of a Regulatory Framework. A typical regulatory framework should consist of:

1. Definition of the different roles assigned to the national entities related to the sector: ministry of health, ministry of environment, operators and regulatory agency.
2. Establishment of realistic planning targets in water quality and quantity.
3. Establishment of national standards to guarantee water quality and quantity for all the users including the poor segment.
4. Development of mechanisms to allow incorporation of PSP
5. Introduction of mechanisms for the approval of water rates.
6. Establishment of a transparent mechanism to subsidize water rates to benefit the poor.

Creation of a strong and capable regulatory agency. Principal requisites are:

1. Mechanism for financing the entity
2. Hiring of very competent personnel for the agency.
3. Mechanism to inform the users about the situation of the system and for preemptive measures in case of emergencies. Methodology to submit complains.
4. Penalties for lack of compliance with the standard.
5. Efficiency criteria including parameters to measure the efficiency of the PSP and others.
6. Procedures to ensure the water rights.

7. Mechanism for an arbiter in case of conflicts between the operator and the users.
8. Preparation of the tender documents for concession contracts.

Chapter Six

Characteristics of Private Sector Participation

Characteristics of Private Sector Participation

The Privatization Trend

Governments in a number of countries around the world increasingly are choosing to privatize the provision of infrastructure services, including the water sector. France and England are leaders and good examples of what can be done when there is a political will.

Two powerful considerations in the decision to privatize are the inability of the public sector to continue funding the projected investment requirements and the belief that the private sector is able to manage and operate more efficiently than the public sector.

The second key driver of the trend toward privatization is the fact that the public sector operators of water utilities manage their systems inefficiently. Unaccounted for water (UFW) is a good indicator of such inefficiency. In most of the countries of Latin America this value is about 50%. This number includes the physical losses (pipes and other appurtenances of the system not properly maintained), free water, commercial losses and pilferage. The lack of universal metering makes it more difficult to ascertain the actual figure but studies demonstrate the high figures in many systems. It is unbelievable that in an important industry like water, 50% of the production is lost. In well-managed systems in developed countries are usually about 10 to 15%.

Finally, the ratio of 10 to 20 employees for 1000 water connections is another indicator of the same problem. In well managed water companies this ratio is about 2 to 3. In the USA the ratio average is 2.7.

When countries in Latin America decide to partially or totally privatize their water supplies and sewerage services, they benefit from two advantages that the private sector can offer. Firstly, efficiency in the delivery of the services is improved due to the incentive for profit in the private sector. Secondly, the private sector is responsible for financing the necessary works based on the revenues paid by users. In Chapter Five, information about the current PSP models in Latin America is presented.

In a public invitation for a concession contract recently in Panama, the government requested the potential concessionaire to finance the investment program and to pay for the right to be the concessionaire as well. This advance payment is called "the canon." The canon would go to the national budget once the contract is signed. This requisite of paying canons is common in other sectors like energy, telecommunications and gambling but not in the water sector.

To obtain the benefits of PSP participation, the political will to reform existing legislation in order to separate three functions that traditionally are handled by the public sector is mandatory: planning, operation and regulation. This sector reform generally requires a decision at a high level of the government. It is usually necessary to pass a law that allows PSP. Sometimes, the necessary legislation is difficult to obtain when the president does not have a majority in congress. In addition, there is an associated trauma in the minds of the industry workers that private intervention means massive lay off, which is generally likely. In most cases, this feeling of removal causes a serious social problem that the governments want to avoid for political reasons. The results obtained in a national referendum in Uruguay a few years ago illustrated this problem. However, implementing PSP might not mean that the existing personnel are adversely affected since the private company needs local support and may carry staff over and/or hire locally as well. Privatization means intervention of the private sector and has different modalities as set forth below.

For any PSP alternative, a sound technical document should be prepared giving priority to system rehabilitation. To start the process to attract private capital it is necessary to assign a public entity as responsible to carry out the task, one with a "pro-investment" attitude. To invite interested parties it is necessary to prepare some basic documentation. It is common today to contract with the so called "investment banks" instead of a consulting firm to prepare an "Informative Memorandum" and to conduct a "Road Show," visiting several countries looking for potential investors and operators. In our experience, the investment banks concentrate their efforts on the financial analysis of the utility and overlook other important aspects of the transaction. In effect, it would be desirable that an independent study be made of the technical and environmental aspects as well as the financial. As an example, in a privatization process initiated in Panama, the following list of issues requiring definition were found during the review of the Information Memorandum prepared by an investment bank.

Outline of a Privatization Contract

- Term
- Inventory of assets
- Estimate cost of the inventory
- Water rights
- Decision about the procedures at the end of the concession
- Review of current legislation related to the sector
- Identification of labor laws affecting the concession
- Characteristics of bulk water contracts
- Methodology of award of the concession

Outline of the Technical Requirements for Operation of Water and Sewerage Services

- Influent water quality
- Effluent standards
- Water pressure
- Interruption of services
- Alternative sources of supply during emergencies
- Flooding and emergencies
- Wastewater treatment
- Public relations
- Responsibility for meeting standard set by the regulating agency
- Economic penalties for non-compliance
- Investment program by stages
- Environmental audit of the system

- Causes for contract cancellation
- Rule for participation of partners

OUTLINE OF THE REQUIREMENTS FOR A REGULATORY FRAMEWORK

- Water rates
- Service quality
- Environmental quality
- Effluent standards
- Rate/Tariff study
- Interim rate structure(s) (5 years)
- Subsidies
- Taxes
- Implementation of the water law(s)

TYPES AND MODES OF PRIVATE SECTOR PARTICIPATION

Since late 1980's many papers have been written about the subject. The author consulted several publications among them those numbered (39) to (48). In addition to the useful information contained in those publications, the author acquired a lot of experience working directly in several privatization exercises. The more relevant are: Argentina (Buenos Aires and Tucuman), Peru (Lima), Uruguay (the whole country), Panama (the whole country), Jamaica (the whole country).

Worldwide experience with PSP in the water sector indicates that there is no single model for structuring private sector management and capital participation that is best. The suitability of a particular arrangement depends on the local institutional capacity and specific project related factors. The different models represent a continuum of allocation of responsibilities and risks between the public and private sector.

As each of the different models is discussed below, a word of caution is offered in making comparisons since the scope of the work involved can vary greatly across the contract types. Besides, there is always a political factor involved in the decision-making process. The table at the end of the section is included as a summary of the variables involved in every modality.

Service contract

In this modality the private sector does not have anything to lose because there is little risk involved. From the users' standpoint, this is not a valid option because there is usually a trade-off between having better service and paying higher rates. On the other hand, from the government standpoint this modality is inconvenient because the government retains the commercial as well as the economic risks of the operation. Furthermore, this modality is relatively easy to cancel and is exposed to the possibility of frequent changes of contractors because the contract terms are usually short-term (1 to 3 years) and the government deficiencies are easily translated to the contractor. Frequently the service contracts relate better to components of the system rather than to the whole.

From the private sector standpoint this scheme is acceptable because there is no economic risk involved. However, the private sector is still vulnerable because of the short-term nature of the contracts.

Management contract

In essence, a management contract is a service contract but for the whole system rather than a portion. Hence, even though financial investments on the part of the management contractor are small the commercial and economic risks still remain on the government's shoulders. In some cases through special arrangements, the government can translate part of the financial risks to the management contractor.

The private contractor typically accepts performance-based fees, which are generally based on physical parameters such as the volume of water treated and achievement of environmental quality standards. The contractor may also bear the risk of legal liability for failure to meet environmental standards. The contractor does not take on the investment and financing risk. The duration of management contracts is generally less than 10 years.

Because they may not require rate increases or significant downsizing of staff, short-term management contracts may be politically more acceptable than forms of PSP such as concession contracts which require cost recovery. Tucuman in Argentina, Cartagena in Colombia and Monagas in Venezuela are some examples of this modality.

Affermage (Leasing)

In France, delegated management has been practiced for about a century[41]. Local authorities can choose either to run their water services directly or to commission this task to a private operator. This decision is made either by the local council or by a water board. The system works in one of two ways: leasing (affermage) or concession. With the affermage option, the local authority bears the initial investment costs and commissions the management tasks to the private operator over a period that can run from 5 to 20 years. The operator is paid for water supplied but must also pay the local authorities rental on the plant it uses. The local authority uses this rental revenue to repay the technical and financial cost on the initial investment. The other option is the concession that is discussed later.

In France, for a total of 36,763 cities there are 15,244 separate supply systems and 11,992 drainage systems (wastewater and storm water). At the present time, concession arrangements cover 75% of the total volume of water supplied. Delegated management is becoming increasingly popular owing to the scarcity of state funding and the need for a substantial initial investment.

In Latin America this modality is very attractive to the private sector because it is only responsible for some commercial activities like maintenance, installation of water meters, construction of small portions of the network. From the users' standpoint this modality is attractive too because they perceive some improvements even though the water rate usually increases. As indicated above, the government or state retains the responsibility to make large capital investments including those for corrective maintenance. This kind of contract can be either medium or long range and it is easily renewed.

Full Concession Contracts

In a full utility concession, a contract allocates to the private sector all the responsibilities of leasing plus capital investment for rehabilitation and expansion.. Because it bears the consequences of both operations and investment decisions, a concession holder faces stronger incentives to integrate and perform both of these tasks efficiently. Greater shouldering of operational responsibilities and stronger incentives means that such concessions offer a much broader scope for productivity improvements through, for example, water loss reduction, lower overheads and more efficient billing and collection.

The concessionaire is remunerated according to contractually established rates (often the bid price in a competitively awarded concession) collected directly from the consumers. At the end of the concession, which generally lasts 25 to 30 years, or long enough to recoup capital investments, the fixed assets are returned to the government and the concessionaire is compensated for the residual value of the facilities it has financed. Although the contract represents the main instrument for governing the concession, a regulatory process is typically developed in parallel to deal with evolving policy concerns and re-negotiation of the contract to adjust rates and assess the performance of the investment program. Thus, because of its impact on a concession's financial viability, regulatory risk must be adequately identified and minimized.

This modality is entirely different from the previous ones because the concessionaire assumes the full financial risks and usually includes most of the economic capital investment risk (sometimes shared with the government). This scheme is attractive to the private sector provided that: a) the concession term is at least sufficient to recover the investment, and, b) the water rates cover the O&M costs as well as a reasonable return on the investment. From the government standpoint this modality is acceptable because most of the economic risk rests with the concessionaire.

There are currently two problems governments encounter related to deciding in favor of PSP:

1. keeping the existing water rate for the first 3 to 5 years to gain user acceptance;
2. solving the social problem caused by the potential removal of the majority of the current personnel. In the Buenos Aires concession, 10% of the shares of the new company were offered to the personnel as an incentive to retire. In this regard, there is plenty of room to negotiate satisfactory solutions to all the actors involved.

A successful concession must guarantee that even though it becomes a monopoly, it will always award contracts to the private sector by the way of public bidding for design, construction, operation, BOT, purchase of components of the system within its service area and beyond. The concessionaire, should be obliged by contract to let other private companies bid. This scheme is adopted in the Chile and in Panama water concessions. The regulatory entity is responsible to assure compliance.

Assets Sales

This modality has been utilized in the water sector in Europe. England is a relevant example. As far as is known this modality has not been applied in Latin America. Some characteristics of the model are presented here.

Assigning the private sector full financial risk and transferring water and sanitation assets to private ownership offers the greatest potential for efficiency improvements and the longest time horizon for investors. In the full privatization model, all assets, operations and future capital investment requirements are turned over to the private sector for perpetuity. To protect consumers from monopolistic behavior, a private owner of a water company generally operates under a license from the government that sets certain service levels and fees/rates. This licensing function may be implemented by a regulatory agency to monitor and enforce service and tariffs. Hence, to protect investors and operators, a predictable regulatory structure is crucial.

Unlike the other contractual agreements examined, private asset ownership provides the opportunity for capital market monitoring. For exchange-listed water and sanitation companies, an important mechanism for disciplining management is the information provided in stock/security prices. Low stock prices can lead to corporate takeovers of poorly performing companies. Also, where a credible regulatory framework exists, asset ownership provides security and collateral for long term investment.

As mentioned above assets sales have been used only rarely. Many countries have explicit laws or constitutional clauses prohibiting private ownership of water assets. Concern is often voiced about the loss of competitive discipline that is maintained in the concession contracts through periodic re-bidding of concession rights.

Build Operate and Transfer (BOT) and Build Own Operate and Transfer (BOOT)

These two modalities are commonly used for new components that are added to the systems. For instance, the incorporation of a new water source by a BOT contract can be used by the concessionaire in a bulk water contract. Currently these contracts are for 10 to 15 years and are very popular in countries like the USA. There is a tendency in Latin America toward this modality because the bidding process is simple and the components can be built and put in operation in short periods of time. In Lima, a new water source (Chillon River) has been contracted by way of a BOT to serve part of the water system. In Panama, the same model was adopted to solve water scarcity in a very dense area not served by the city water system.

Franchising[48]

The use of franchising[48] is well known in other businesses like the food industry, hotels, post offices, etc. The seller of the franchise is the franchisor and the operator is the franchisee who receives the know-how from the franchisor. This kind of arrangement is similar to a management contract with an international operator who supports the local operator by providing technical assistance, skilled personnel, and technical advice for the purchase of special equipment.

In the water industry this modality per se is not being utilized. A study conducted by the World Bank ends saying: *"given the potential benefits that may be obtained through the application of franchising to the water sector, it is recommended that the issue be further investigated and discussed."*

Public-Private Partnership

In general, public-private partnerships[44] describe a contract relationship whereby the public sector entity remains the owner of the assets, controls management of the assets through the contract, sets rates and provides strategic long term planning. The private contractor simply becomes the executor of public policy, a contracted extension of public resources planning and management, and one element of the overall municipal budgeting process. This modality is very common in the USA and now becoming popular in Latin America for small municipalities.

Under this scheme in some countries in Latin America, the municipality creates a municipally owned corporation offering a little less than 50% of shares to the private sector (thus retaining ownership for the government). Once the public corporation is operational, it invites interested operating contractors to submit proposals to delegated management. The municipality guarantees the water rates to cover operating expenses and a reasonable profit for the operator.

The Water Partnership Council, a Washington, D.C. based association of private management services companies recently released a handbook for the public sector, *"Establishing Public/Private Partnerships for Water and Wastewater Systems,"* that presents the array of possible service relationship with the private sector. Covered are: evolving role risk transfer and performance, financial guarantees offered, measurable benefits and a guideline to the "dos and don'ts" of soliciting and structuring successful partnerships that best meet the expectation of the communities. Also, the World Bank developed a useful tool called *"The Infrastructure Privatization- Nationalization Wheel and Ten Questions for a Sustainable Regulatory Regime for Infrastructure."*

The American Water Works Association also recently published software called: *"Balance Evaluation of Public-Private Partnerships (the 3 'Ps'),"* to facilitate the understanding of the different scenarios and decision-making toward privatization. This program takes into account the relevant factors in a systematic approach to obtain the best-balanced option for privatization. The parameters are: information about the partners, technical, environmental and political aspects, risk factors. The model makes explicit the ranking of the options, education of the partners, and information to defend option selection using a logical process.

The following Table presents a summary of the PSP alternatives presented and discussed above.

PSP Summary

Alternative	Ownership	Financing	Management
Service contract	Public	Public	Private
Management contract	Public	Public/Minor Private	Private
Lease/Affermage	Public	Public/Minor Private	Private
Concession Contract	Public	Private	Private
Assets Sales contract	Private	Private	Private
BOO, BOT, BOOT contract	Private/Public	Private	Private
Franchising contract	Private	Private	Private
Public- Private Partnership	Public	Public/Private	Private

Chapter Seven

Conclusions and Recommendations

Conclusions and Recommendations

Conclusions

1. Evaluation of water and wastewater services in Latin America indicates that at the end of year 2000, there was a large deficit of infrastructure, requiring investments of the order of US$ 100 billion to reduce the current deficit by half in the new decade of water period 2005- 2015. Political decisions by the governments to adhere to the new decade recommendations are instrumental. In the past, and in two similar decades launched by the United Nations, the response of the governments was insufficient. *Water for All* is not an easy achievement and becomes more difficult due to the requirements established by Agenda 21.
2. Given the hefty investment required for the new water decade and the economic and social situation of the countries, more funding from the private sector is needed. Most Latin American countries realize that there is no other practical option for funding but to count on the private sector. Economic and government stability of the countries is essential to attract the private sector. Attracting private capital also requires contracts and security agreements for a potentially secure revenue stream.
3. Before a national investment program is designed under the decade's recommendation, the most cost-effective projects should be proposed that comply with the economic requirements of having a positive net present value and an economic internal rate of return over 12% using a discount rate of 12%. Furthermore, all projects should conform to the policies and standards established by national and international funding organizations. The appraisal of investment projects for

funding requires the participation of a specialized multidisciplinary team of professionals to determine the feasibility and sustainability of the projects. Technical, environmental, institutional, financial, socioeconomic. legal and political aspects should be reviewed in detail.

4. The majority of water sector deficiencies are of an institutional nature: poor management, inadequate maintenance practices, high levels of UFW, insufficient cash flows due to low tariff levels, poor collection, lack of proper communication systems and instrumentation, non transparent subsidies, redundant public utility personnel and absence of strategic planning. Hence, sector reform is essential. Recommendations to the countries by the IADB, World Bank and other agencies to implement sector reform have been given as a way to discipline the sector. Technical assistance from those and other international entities is always available.
5. The sector reform must separate planning, operation and regulation functions. Planning rests often with the health ministries. However, planning residing in the economy ministries would expedite the incorporation of projects into national economic planning and budget processes, as it is in Uruguay. A strong and capable regulating agency is basic to sector reform. Local operators, duly trained, are encouraged.
6. Currently, there are several institutional models in place in Latin America with an incipient trend to privatization. Governments realize that private sector participation is essential for a sustainable development of the sector.
7. The current institutional models do not include provisions and standards to serve the low-income people located in urban areas. Several cost-effective solutions using appropriate technology should be incorporated into the current standards in order to meet this need. The utility must play a role in implementing simplified systems and appropriate technology including taking responsibility for changing design parameters, supporting local initiatives, and encouraging private contracts for design, construction and operation. For the poor connected to the urban water systems, the water rate structure established by the utility should be applied. When the low-income

users cannot afford to pay the water rate a subsidy mechanism should be implemented. Often the poor pay for relative luxuries but are reluctant to pay for water.

8. The organization of the low-income communities in cooperatives has proven to being a valid model by incorporating the users in the decision-making process. The central utility, whether public or private, should incorporate the proper mechanism to give assistance and training to the cooperatives even if they are not located within its service area. It should bear in mind that they are potential clients in the future.
9. Several privatization options are available. There is no model that might be considered universally applicable. The adoption of a particular model depends on the particular situation. The main considerations to facilitate the private sector participation are:
    a. Government political and financial commitment.
    b. Contractual and regulatory structures that minimize the uncertainty and provide flexibility in re-negotiation and in operational autonomy.
    c. Competitive tendering is an important tool with which to generate information on assets, environmental profiles and tariff levels. The investment banks and consulting firms have the necessary experience and knowledge to conduct evaluations of the privatization process.
    d. Concessions and asset sales provide the broadest scope for operational and financial improvements. However, the asset sales approach has not been used in Latin America.
    e. When concessions and asset sales are not politically acceptable, delegated management (public- private) is applicable in the transitional period to prepare the entity for future full privatization.

Recommendations

Introduction

The launching by The United Nations of a New Decade (2005- 2015), to expand and enhance the water and wastewater sector in Latin America, is a golden opportunity for Latin America's countries to work hard to reduce the existing gap of services for *All*, including the low income sectors of the populations. As described herein, there are many variables to consider for successfully achieving the target established by the Decade.

LIST OF RECOMMENDATIONS (DECALOGUE)

To the Latin America countries. Endorse and support the development of a new era, in the third millenium, to provide basic water and wastewater services to All. The year 2005 is the preparatory year of the decade. Hence, it is recommended that in every Latin America country, an inter-sector steering national committee be created within the economy ministries as responsible entities to implement the commitment. Public and private interests should be duly represented on the committees.

To the Pan American Health Organization. Should help the countries in the identification and project preparation of their national plans and programs, and related activities through its headquarters, CEPIS, ECO and country offices.

To the international development banks and funding agencies. World Bank, and in particular, the Inter American Development Bank as a regional bank, should support the countries' efforts by providing banking resources, including partial financing for concessions, and technical cooperation.

To the more developed countries of the region. Provide technical assistance (Intra) to the less developed countries as a way to transfer Latin America's technologies and experience on the best practices.

To AID (USA), CEMAT (Guatemala), DANIDA (Netherlands), GTZ (Germany), IDRC (Canada), IIC/JICA (Japan), UNDP (United Nations), USEPA (USA), and WHO/UNEP (World Health Organization). As international agencies devoted to the water and wastewater sector, provide technical assistance and to conduct research programs to benefit the water and wastewater sector in Latin American.

To the American Water Works Association. Continue promoting technical activities related to the water and wastewater sector in Latin

America through its International Affairs Committees and invite the Water Environment Federation to jointly develop that effort.

To the Inter American Association of Sanitary Engineering and Environmental Sciences (AIDIS). Continue supporting the countries' plans and programs, through its headquarter and chapter offices in every country of Latin America, by a) dissemination of technical publications and results of research studies, b) creation and implementation of a roster of international and national consultants, as a basic resource, to help the water and wastewater sector development.

To the water industry. Incorporate Latin American technologies for/in: a) project design (consultants), b) operation of the services (operators), c) training of local personnel, and d) manufacture of its products.

To the Latin America's universities. Should actively participate in programs to: a) develop new technologies, b) enhance the formal education with due attention to the needs of the water and wastewater sector, and c) train local operators.

To Latin America's decentralized political organizations such as states, departments, provinces and municipalities. Should play an active role in: a) promoting community participation, education and training to facilitate acceptance to the necessary water rates to finance their plans and programs, b) demanding the creation of strong and capable regulating agencies to preserve the quality of the services and the environment, c) requesting that the eventual concessionaires and operators of the water and wastewater utilities extend their service areas to incorporate the scattered population using appropriate technological solutions, and, d) establishing necessary subsidies to help the most needed sector of the population to accede to the water and wastewater services.

Appendix I

Partial List of Projects Directed by the Author at the IADB

Brief Description of the IADB's Projects

Barbados-South Coast Sewerage Project

The project's main objective is to clean up the near shore waters of the South Coast of Barbados, to reduce the bacterial contamination to accepted international standards, to protect the reefs, marine life and water quality of the bathing beaches. Population to be served by the system, 52,000 people; cost of the project US $73.1 million. Principal components of the project are: (a) 40 km of collection system, from 8" to 30" diameter, (b) five lift stations, (c) a main pumping station as part of a primary wastewater treatment plant with fine screenings of 27,930 m^3/day, including ozone equipment for odor control (d) 1.1 km of ocean outfall, 26" diameter (e) service connections and in house connections, and (f) institutional strengthening of the Barbados Water Authority. Very detailed marine studies were carried out, and mathematical models were used to estimate the initial dilution, dispersion and bacterial die-off to design the ocean outfall.

Nicaragua-National Water Supply and Sewerage Rehabilitation Program

The program's main purpose was to rehabilitate and optimize the existing infrastructure in 27 minor cities of over 5,000 people and Managua, the capital city over 1 million people, by improving of their operation and maintenance, drinking water quality, quality of the receiving waters and executing a rehabilitation component. The program includes also the improvement of the commercial activities of the National Water and Sewerage Institute (INAA), a marginal cost tariff study and technical assistance activities. Total cost of the project US $64.2 million.

Argentina-National Water Supply and Sewerage Credit Program

IADB and the World Bank, joined efforts to partially finance an innovative credit program to solve the critical environmental conditions of the sector and strengthen the capacity of the provincial and municipal entities to self finance their future investment program. Every national entity has the right to request from Consejo Federal de Agua Potable y Saneamiento (CoFAPyS), a loan to develop a very orderly project in the following sequence: (a) institutional improvement, (b) rehabilitation, and once the two previous stages are complied (c) the expansion of their systems. Total cost of the program is US$ 250 million.

In addition to this program which was approved in 1991, Mr. Alfaro had been the team leader of two previous projects in the great Buenos Aires to provide water supply to 2 million low income people in the municipalities of la Matanza, Moron and 3 de Febrero. These areas are now under the operation of a concessionaire. The total cost of these 2 projects is US$ 450 million. Mr. Alfaro also participated on the Buenos Aires Water Supply and Sewerage privatization which ended with the concession to a big international consorcio leaded by Lyonesse dex Eau.

Venezuela-Water Supply for the Central Region

Within the overall objective of a comprehensive environmental Master Plan of the Lake Valencia watershed, it was defined as a first stage, to develop a water supply project to provide drinking water for 3 million people living in Valencia, Maracay and several small towns. The project comprises: (a) a new intake at Pao River for 12 m^3/sec of capacity, (b) two pumping stations, (c) a transmission line of 72 km of 2.1m diameter (d) a conventional water treatment plant in two stages of 6.0 m^3/sec each one, (e) a gravity line around the lake to supply Valencia and Maracay of 80 km and diameters ranging from 0.9 to 1.4m, (f) a rehabilitation component for the small sub systems, (g) an unaccounted for water study and (h) an environmental plan to protect the watershed.

The project also included the creation and implementation of a New Authority to serve the area. Total cost of the project is US$ 384 million.

The second stage of the program comprises a project to control the heavy pollution of the lake, by intercepting the several discharges of raw wastewater and treating them in three secondary treatment plants. Total cost of the project is US$ 130 million.

Paraguay-Asunción Water Supply Expansion Project

The main objective of this project was to provide water supply to low income people living in 5 municipalities located in the Asuncion Metropolitan area. The total benefited population is 1.5 million. The project comprises: (a) a new intake and pumping station at the Paraguay River with 4.0 m^3/sec of capacity, (b) a patented water treatment plant of 1.3m^3/sec of capacity, (c) 4 pumping stations, (d) 5 elevated reservoirs, (e) 30 km of transmission lines with diameters ranging from 16" to 32", (f) 170 km of distribution network with diameters ranging from 6" to 32", and (g) 25,000 service connections including water meters. Asuncion system is one of the few in Latin America with 100% metering. The total cost of the project is US$ 75.8.

Mexico-Tijuana Water Supply and Wastewater Project.

The objective of this project was to rehabilitate and to expand the water supply and sewerage systems of Tijuana City to properly attend the demand of 1.2 million people. The water supply system comprises: (a) 33 km of transmission lines with diameters ranging from 6" to 16", (b) rehabilitation of the transmission line from La Misión to Tijuana with 20" diameter, (c) 104,000 m^3 of the reserve water in 10 reservoirs, (d) 3 pumping stations, (e) 95 km of distribution network with diameters ranging from 4" to 8" and (f) installation of 50,000 water meters. The sewerage project comprises: (a) 36 km of interceptors with diameters ranging from 24" to 96", (b) 73 km of collectors with diameters ranging from 8" to 36" and (c) 49 km of minor collection system with diameters ranging from 8" to 10". It was agreed on between the Mexican and the USA government to build a binational wastewater plant with secondary treatment in the San Diego area to treat about 50% of the wastewater from the service area. The other 5% of the wastewater is treated in a secondary plant located in the Mexican side. The final disposal of both effluents is going to the sea and it is in compliance with the California's Water Quality Standards. Total cost of the project is US$ 91 million.

Uruguay-Montevideo Sewerage Project

For more than 35 years the Montevideo beaches, which are a major tourist attraction of the city, suffered of a heavy domestic and industrial pollution. The Comprehensive Master Plan prepared in 1972 indicates that the final solution is to intercept more than 20 raw sewage discharges, to convey the sewage to a preliminary treatment plant and to dispose it into the river via an underwater outfall. It was also recommended to build the interceptor system in two stages, one serving the east part of the shoreline, and the other one servicing the west part but having a single final disposal point at Punta Carretas. Both stages comprise the following: (a) 20 km of interceptors with circular pipes and special sections to be build on site with sufficient capacity to drainage the combined wastewater effluents, (b) 6 lift stations, (c) a preliminary treatment plant to eliminate floating solids, sand and grease, (d) an underwater outfall of 2.1 km of 1.8m diameter and (e) 180 km of collection system to attend the demand of several unsewered sections of the city. The total population benefited by this project is 1.3 million people. The technical design of the underwater outfall required a detailed water quality study and the use of very advance mathematical models to simulate the difficult physical conditions of the several disposal points considered in the alternatives study.

Barbados-Comprehensive Solid Waste Management Program

Bridgetown the capital city of the country, suffers a serious environmental problem due to the inadequate handling and disposal of the domestic, commercial and industrial solid wastes. The existing facilities do not have sufficient capacity particularly, the sanitary landfill, to handle the ever increasing amount of solids. Besides, the existing Authority does not have the required technical capability and the necessary funds to carry out an immediate project to improving the system. Mr. Alfaro, has worked in the conceptual design of a Comprehensive Program and a Technical Assistance has been granted by IADB to carry out a feasibility study and the design of the priority works, to submit a loan request for the first stage of the program. It is estimated an investment of US$ 30 million to build the facilities.

Appendix II

Scope of Work UFW (Non Revenue Water) Study
Cuenca's Water System
By David Goff

Proposed Scope of Work

I. VERIFICATION AND ORGANIZATIONAL TASKS

A. In order to effectively complete the proposed Scope of Work, the Consultant will participate with/engage responsible staff in practically all departments within ETAPA (Local Utility). To achieve the most benefit from this contract, it is recommended that ETAPA appoint a qualified liaison who will work alongside with the Consultant. The liaison will be responsible for recording, monitoring, becoming familiar with and understanding the work and processes utilized by the Consultant, so that ETAPA can periodically repeat any or all of the work and investigations.

In order to identify all sources of UFW, the consultant will collect and review all relevant existing data and determine its source, validity and reliability, paper-train and frequency of reporting. He will perform interviews with responsible staff, and inspect all relevant facilities and equipment.

B. An Audit year will be identified; e.g., 2001, and the data will be entered and organized into the format of an Audit spreadsheet which will then be adapted for use as a monthly report for monitoring UFW. The normal adjustments will be made to the Audit data, including Meter-read Lag and any necessary adjustments, which are a function of the monitoring and reporting frequency of Production and Tank Levels.

C. Review the existing Divisions in ETAPA and their relevance and significance to the UFWP (UFW Program). Develop both the management and technical hierarchy and reorganizational schematic for ETAPA's UFWP; which includes a 'point'/strategic committee.

D. Complete the specifications for the Meter Warehouse, the water meters and the bench-testing equipment. Review the existing floorplans of the Warehouse. Recommend specifications including meter size(s), simultaneous meter-testing capacity, flowrate range capacity and brand for the bench-testing equipment. Also recommend specifications for comparative meter testing equipment, which at least includes size, brand, capacity for flowrate ranges, automatic recalibration and digital, electronic-reset registers. The comparative testers will/can be used for both the larger commercial, mastermeters and hydrant/ construction turbine meters. Smaller comparative testers with similar specifications should also be purchased for on-site, in-place comparative testing of the smaller residential meters. All testing equipment should at least meet AWWA requirements and other Latin American specifications.

E. Analysis of the commercial area of ETAPA, to determine the convenience of outsourcing this component. In this area, the situation of the users cadaster and connections, the methodology used for billing and collection, the organization, the water rates, the tariff structure, the UFW levels present and the forecast for future situation once the UFWP is implemented, the tariff evolution, the legal aspects, and macro and micro metering, will be appraised. To appreciate the evolution of the macro and micro metering, a comparison will be made with the indicators of other Latin America working at optimal level.

II. STRATEGIC INVESTIGATIONS TO DETERMINE THE RELATIVE SIGNIFICANCE OF THE DIFFERENT SOURCES OF UFW

Initial objective is to identify, verify and use "Best Available Data" to best estimate the relative significance of ***"Apparent versus Real Losses"*** in the UFW percentage. ***Apparent losses*** are errors and "paper"' losses, which in most cases, when corrected and/or reduced do not reduce production requirements, but recover significant unmetered, unbilled, revenue. ***Real losses*** are leakage. This will enable ETAPA to confirm active programs and proposed plans, as well as to prioritize and approve recommended programs and investigations to reduce UFW in the short and long term.

A. Apparent Losses

1. Production Data, Master Meters and Storage Tanks
   a. Test "finished water" and/or any other mastermeters used to develop the specific Production Data used to calculate the UFW percentage.
   b. Review the efficiency of conserving water resources practiced through the production, treatment and storage process; e.g., recycling of backwash and sludge cake waters, unmetered operation & maintenance uses.
   c. Review records to determine potential for Leakage from Storage Tanks.
2. Consumption Data
   a. Identify and verify billing and meter-reading adjustments performed; edits, credits/debits. Identify and verify reporting and work-order generation regarding meter and read problems. As it has already begun with investigating commercial accounts that are <20m^3/mes; compare and further investigate the average monthly consumption per account for the different classes. E.g., as the table below shows, (adapted from ETAPA's information), the average monthly consumption of "Special" accounts is the highest, and "Tariff 50% (E)" accounts only pay for half; yet consume twice as much as "D".

METERED COMSUMPTION	Consumption m³/mo.	% of Total Consumption	Number of Users	Non functioning meters	Stopped Meters	Working Meters	m³/mo per/ meter
C:Construction	6,260	0.3%	538	83	149	306	20
D:Special with Meter	61,931	3.4%	194	11	11	172	360
E: D with tariff 50%	41,261	2.2%	66	6	1	59	699
I: Industrial with Meter	76,720	4.2%	255	18	1	236	325
M: Commercial with Meter	146,686	7.9%	2,629	179	149	2,301	64
N: Residential with Meter	1,512,594	82.0%	49,590	3,388	7,869	38,333	39
Total	**1,845,452**	**100.0%**	**53,272**	**3,685**	**8,180**	**41,407**	

Note: There are 391 users without meters.

b. Identify and estimate sources of apparent losses responsible for UFW, such as construction, unmetered, metered unbilled, stopped, nonfunctioning, inaccessible and illegally connected.

c. Unregistered Consumption: Design series of investigations to estimate the various sources of unregistered consumption shown on Table before, which amount to 23% of all metered connections/services. Note that in many systems there is significant lowflow losses from users that can be below the detectable limit (BDL) of the customer's meter. This may be from household plumbing leaks, or from "lowflow" filling of the cubic meter storage tanks installed in most residences.

d. Registration Accuracy: Reducing meter error will probably produce the most significant reduction in UFW and recover significant unbilled revenue. Investigate Cuenca's meter inventory, review, verify and develop data and other requirements in order to determine how best to design the statistically-valid, representative samples for accuracy testing.

- How many must be tested in order to determine optimum meter life, and when and which to replace? Continue to determine increased consumption and revenue from meter replacement program need to compare and eliminate any seasonal or other consumption variation.
- Does the number of medium-sized meters warrant samples, as well as, the Residential (small ≤1") and commercial/industrial (large ≥2") meters? Bench-testing or on-site, comparative testing for residential meters?

- After reviewing ETAPA's distribution and other analyses of existing meter test results, shown before there can be significant difference in the calculated average meter accuracy, depending on whether the meter accuracy at the three flow rates are numerically averaged or weighted. "Weighting" should be based on estimated percentages of consumption at the three flow rates. Low flowrate meter accuracy is generally the first to deteriorate, but usually only accounts for less than 15% of normal consumption; versus 33% using a simple average of three flowrates. Therefore, differences between using simple and weighted averages can sometimes be >50%...which has very significant impact when extrapolating results from samples to an entire meter inventory. How best to weight the averages of the accuracy at the 50, 250 & 1.500 liter/hour flow rates used for meter testing? Use estimated percentages of flows to correct the current projections of existing meter test datasets; install dataloggers.

B. REAL LOSSES

1. *District Analyses.* This process should be evaluated to determine how to establish a permanent system to monitor and control both consumption, losses and UFW.
   a. ETAPA has eighteen (18) existing districts, each with an independent distribution system. Each is also reported as mastermetered; however, by metering the flow into each of the 18 respective storage tanks.
   b. Each must be verified to be effectively isolated; e.g., with properly closed valves; metered emergency interconnections.
   c. Also, in order to determine "minimum night ratios" and estimates of suspected leakage, as well as to prioritize the districts with the most water losses, the consumption of the customers with significant 24-hour flows must be identified and accurately metered.

III. FIELD INVESTIGATIONS TO CONTROL UFW

A. Production/MasterMeter and Consumption Meter Testing

Based on inspections and investigations of the meter inventory and maintenance records, develop the statistically valid, representative samples and recommendations and perform testing using the appropriate comparative or bench-testing for the respective representative meterset samples.

B. Tank Inspections

Based on available tank operation and maintenance records and results from previous inspections, recommend required tank inspections for leakage; where possible, perform leakage tests by isolating tank and monitoring level changes.

C. District Analyses

After verifying the integrity of the districts, and any interconnection and 24-hour users monitoring requirements, perform district analyses during the night(s) to determine "minimum night ratios" and estimate "*real loss*" leakage. Investigate and determine how to correct for tank level fluctuations, or relocate mastermeters to tank effluents.

D. Leak Detection

Depending on the ability to perform, and the results of the District Analyses, as well as, progress and results from recent leak detection surveys (250 Km) and the associated length of mains and number of services within the prioritized "Districts". Perform leak detection investigations on either:

- Each of the entire prioritized "Districts"
- Representative samples of the large, prioritized "Districts"; developed using available leak/break repair records and data on age and materials of mains and services, or;
- Statistically valid and representative samples of mains, services and hydrants throughout the entire distribution system, based on available leak/break repair records and data on age and materials of mains and services.

IV. TECHNICAL & ECONOMIC ANALYSES

A. Identify and use the respective rates; $ per cubic meter; and verify the marginal cost of production. Determine relevant operation, maintenance and repair costs, and incorporate any other relevant costs for use in economic analyses.

B. Analyze the various meter test results, determine their adequacy and significance and extrapolate the estimated error for the respective, entire meter populations.

C. Analyze and determine the *Apparent Loss*-UFW reduction and the associated recovered revenue.

D. Analyze the "District Analyses" results to identify which "districts" have significant leakage and prioritize those districts for leak detection surveys.

E. Analyze the Leak Detection results, determine their adequacy and significance, and extrapolate the estimated leakage error for the respective districts and the entire distribution system.

F. Analyze and determine the *Real Loss-UFW* reduction and the savings in marginal costs.

V. RECOMMEND & PRIORITIZE UFWP AND PROGRAM ELEMENTS

A. Enter the calculations and estimates of the UFW associated with the various tasks/investigations onto the Audit Spreadsheet; identify their relative significance and recommend investigations and program improvements for how best to monitor, control and reduce UFW in the short and long term.

B. Prioritize, design, cost and schedule the recommended investigations and improvements; after reviewing the existing ETAPA plans.

C. Identify and develop data required to monitor various, relevant *Performance Indicators*, such as:

- the ***Infrastructure Leakage Index (ILI) = Real Losses/UARL***

Well-managed systems are expected to have low values; near 1.0, while systems with infrastructure management deficiencies will present higher values.

Appendix III

Executive Summary of the Environmental Impact Assessment of Barbados South Coast Sewerage Project

Prepared for IADB by Dr. Larry Canter

1. Project Description

The South Coast Sewerage Project has been proposed by the Government of Barbados to meet the sewerage collection, treatment, and disposal needs for a 3 square miles unsewered area along the South Coast of the Caribbean Island of Barbados (Barbados has a total land area of 166 square miles).

The South Coast area is heavily developed for tourism, including a number of hotels and public and private beaches. The project is needed because of the deteriorating coastal water quality, adverse impacts on near-shore coastal reefs and commercial fisheries, and the resultant increases in beach erosion due to reef deterioration. Longer-term needs for the project include reducing the risk of ground water (sheet water) pollution in the South Coast area and improving the general health and sanitation conditions for all the Barbadians, including those Barbadians living in the area to be sewered, and tourists residing in the South Coast area or visiting South Coast beaches,

The proposed South Coast Sewerage Project includes three key components: 1) sewering of about 3 square miles in a rectangular strip which is about 6 miles in the east-west direction and 0.5 miles in the north-south direction, including four lift stations; 2) primary level sewage treatment plant (based on fine screenings) designed for an average dry weather flow of 9,280 cubic meters per day; and 3) force main from the sewage plant at Graeme Hill to Needham's Point and 1.0 kilometer long outfall line from Neeham's Point to the marine outfall location west of the Carlisle Bay. The construction phase is expected to last 3 to 4 years, and the operational phase for 20 years or more.

2. Environmental Assessment Results

The environmental impact assessment (EIA) study was prepared following the review of several existing reports on the proposed Coast Sewerage Project and related reports dealing with Barbados' water resources, coastal reefs, beach erosion, and ground water pollution, and the operation of the Emmerton sewage treatment plant in central Bridgetown. In addition, the consultant, Dr. Larry Canter of the University of Oklahoma, who prepared the EIA report, visited the Inter American Development Bank (IADB) in Washington, D.C. and the IADB office in Bridgetown. Barbados, in April, 1991. The purpose of the IADB visits was to enable discussions with pertinent officials regarding key environmental issues related to the proposed project. In addition, while in Barbados the consultant had meetings with key staff persons associated with the Greater Bridgetown and South Coast Sewerage Project, the Bellairs Research Institute, the Coastal Conservation Project Unit, the Ministry of Health, the Barbados Water Authority, and the Emmerton sewage treatment plant.

The organizing feature of the EIA report is a simple interaction matrix shown in Table 1. The matrix displays the potential beneficial and detrimental impacts of the project construction and operational phases on nine environmental factors/resources. The key beneficial impacts are related to the operational phase and include: (1) reductions in the pollution risks to the ground water (sheet water) in the South Coast area to be sewered; (2) improvements in coastal water quality (primarily bacteriological and nutrient quality) due to decreased untreated sewage discharges in the South Coast area, with these improvements leading to decreases in the deterioration of near-shore coastal reefs, enhancement of South Coast fisheries, and reductions in beach erosion; and (3) improvements in tourism opportunities due to cleaner beaches and better coastal water quality and coral reefs (for diving).

Interaction Matrix for South Coast Sewerage Project

		Construction Phase				Operation Phase			
Environmental Factor/ Resource	**Existing Quality**	**Collection System**	**Treatment Plant**	**Outfall Line**	**Resultant Quality**	**Collection System**	**Treatment Plant**	**Outfall Line**	**Resultant Quality**
Air quality	In compliance with air quality standards	A/M	A/M	a	Dusts,CO	a (odor at lift stations sites)	A/M	O	localized odor
Noise	Typical of urban residential area	A/M	A/M	a	increase in local noise	a (pumps)	a	a (pumps)	Small increase in noise
Ground water	Satisfactory for area	O	O	O	same as existing	b	b	b	better quality due to less sheet water
Graeme Hall	Natural biological resource	N/A	a/M (no encroachment)	N/A	Some local disturbances recovery expected	N/A	O	N/A	Same as existing
Beach erosion/ coral reef water quality	Erosion of .0.1 to 0.3 m/year deteriorating coral reef and coastal water quality	N/A	NA	a (water quality)	turbidity increase	b	SB	N/A	improve quality
Coastal fisheries	Some decline due to deteriorating coral reef and coastal water quality	N/A	N/A	a	local turbidity	b	SB	N/A	Improve quality
Marine environment at outfall diffuser	Good	N/A	N/A	a	some local disturbance	N/A	N/A	a	small decrease in quality
Traffic	Current problem	SA/M	a	a	increase in congestion	a	a	a	continued problem due to tourism increase
Tourism	Important to economy	a	N/A	a	traffic congestion might cause decrease	B	B	B	increase in economy

A= adverse impact; M= mitigation measure planned for adverse impact; a= small adverse impact; O= no anticipated impact; N/A environmental factor not applicable; SA= significant adverse impact; b= small beneficial impact; B= beneficial impact; SB= significant beneficial impact.

The key negative (adverse) impacts of the proposed South Coast project are primarily related to the construction phase and include: (1) local decreases in air quality due to construction dusts and carbon monoxide emissions from construction vehicles; (2) local increases in noise due to construction activities; (3) some terrestrial and aquatic disturbances in the Graeme Hall area due to construction of the sewage treatment plant; (4) local increases in the turbidity of coastal waters and resultant small impacts on coastal fisheries due to construction of the outfall line; (5) disruptions in South Coast area traffic, particularly along Highway 7, due to project construction activities; and (6) short- term disruption in some South Coast tourism due to construction activities. During the operational phase, the adverse impacts include: (1) localized odors and noises at the four lift stations; (2) possible decrease in air quality in the vicinity of the sewage treatment if the ozonation system and/or screening incinerator are not properly maintained; (3) localized noise in the vicinity of the sewage treatment plant; (4) localized water quality disturbances in the the vicinity of the marine outfall line discharge point; and (5) increases in South Coast traffic congestion due to anticipated tourism increases (these increases are indirectly related to the project).

Several alternatives to the proposed project were identified by the consultants (Reid Crowther) to the Greater Bridgetown and South Coast sewage project. Examples of these alternatives include: (1) use of conventional sewer construction technology (trenching) versus a "No Dig" technology in certain areas to be sewered; (2) four locations for the sewage treatment plant (Drill Hall, Pavillion, Graeme Hall, and Maxwell): (3) primary versus secondary levels of sewage treatment; (4) incineration of plant screenings versus land disposal of screenings; (5) treatment plant enclosure and ozonationsystem for odor control versus no plant enclosure system; and (6) different locations for marine outfall line. Both economic and environmental factors, along with public input, were used in the evaluation of the alternatives and the selection of the proposed South Coast Sewerage Project.

A very careful ecological analysis of the potential Graeme hall site for the sewage treatment plant was conducted prior to its inclusion in the proposed project. In addition, the engineering consultants gave careful consideration to the need for and design of a screening incinerator and odor control system at the sewage treatment plant.

3. Protection and Mitigation Measures

   A number of environmental protection measures as well as mitigation measures for minimizing the adverse environmental impacts of the proposed South Coast Sewerage Project have been identified. These measures include the following:

   a. Routine construction phase mitigation measures should be used for the sewerage system, sewage treatment plant, and force main/ outfall line. Examples of such mitigation measures are included in Appendices I and J of the EIA report. A construction phase mitigation plan sshould be developed by the Greater Bridgetown and South Coast Sewerage project staff (or their consultants), and this plan should be included as part of the construction phase specifications and/or documentation. (The mitigation plan should be approved by the IADB as a loan condition).
   b. Due to the potential reductions in adverse environmental impacts which can occur due to the implementation of the "No Dig" sewer construction technology, usage of this technology in areas decreed appropriate, from an engineering perspective, is recommended.
   c. The proposed sewage treatment plant includes an incinerator for the screenings. Appendix E in the EIA report includes design and operational specifications for the incinerator. If the proposed incinerator is not properly operated and maintained, problems will occur in relation to local air quality deterioration and the build-up of screenings for land disposal. In addition, a solid waste management study is currently underway in Barbados, with the purpose being to identify appropriate land disposal sites and/or incineration technologies for Barbados. Therefore, it is recommended that the purchase of the screenings incinerator be postponed, with the screenings subjected to current land disposal practices until such time that a comprehensive solid waste management scheme is implemented for Barbados. At that time, the screenings might be subjected to an area-wide incinerator for Barbados solid waste.

d. The proposed sewage treatment plant is to be enclosed with an ozonation system to reduce odor emissions. This design and the environmental impact implications are described in Appendices F and G of the EIA report. It is recommended that special attention be given to the operational considerations related to this system. Efforts should be made to provide plant operators with appropriate training, and the sewage treatment plant must be operated in a manner that would allow sufficient data collection to ensure an efficient operation. Reporting of the operational success of the ozonation system. Reporting of the operational success of the ozonation system should be required by IADB until its successful operation is demonstrated.
e. Efforts should be made to develop appropriate restrictions on land usage and encroachment in the Graeme Hall and Western Lake area which is primarily associated with the proposed sewage treatment plant site. This effort will also involve coordination among both governmental and nongovernmental organizations. Due to the importance of this ecological resource in Barbados, efforts should be made to preserve its existence and to allow natural recovery processes to occur.
f. It is also recommended that as part of this project a sufficient spare parts inventory be maintained for both the collection system as well as the sewage treatment plant. In the recent past there have problems due to administrative requirements of actually getting replacement parts in a timely fashion for the Emmerton sewage treatment plan. Accordingly, plans should be developed to preclude this from occurring in the case of the South Coast sewage treatment plant.
g. A positive program should be developed to control the quantity of septage introduced into the proposed South Coast sewage treatment plant. Information on currently effective efforts for septage control at the Emmerton sewage treatment plant would be useful to consider for the South Coast sewage treatment plant.
h. Job descriptions for all employees of the South Coast project should be prepared to facilitate system operation. Again, recent

efforts in this regard for the Emmerton sewage treatment plant are germane to the South Coast plant.

i. Construction contract specifications should delineated the procedures to be followed in the event that previously unknown historic or cultural resources are found during the construction phase of the proposed South Coast sewerage project.

j. An integrated and coordinated environmental monitoring program should be developed by the Government of Barbados. This program should make use of existing monitoring programs and be expanded to ascertain the environmental improvements resulting from the proposed South Coast sewerage project. The program components should include:

    i. coastal water quality monitoring (for point and diffuse sources).
    ii. coastal reef monitoring
    iii. fisheries productivity monitoring, and
    iv. ground water monitoring in selected South Coast wells.

Information from this monitoring program can be used as a basis for future West Coast project. A 10-year monitoring plan, including reporting, should be developed by the Government of Barbados and approved by the IADB as a loan condition.

References

1. J. Alfaro. Tres Decadas de Saneamiento en Latinoamerica. CEP International Inc. Maryland. USA. 1999
2. J. Alfaro. IADB's Activities in the Water and Wastewater Sector. Twelve Papers. IADB. Washington, DC.USA.1989
3. E. Rojas, J. Alfaro. Principal Environmental Problems in Mega Cities. IADB. Washington, DC. USA. 1989
4. J. Alfaro. Position Paper on the Decade of Water and Sanitation 1980-1990. IADB. Washington, DC. USA. 1980
5. J. Alfaro. Maintenance of Urban Water Supply and Wastewater Systems. IADB. Washington, DC. USA. 1985.
6. PAHO. Regional Report on the Sector Evaluation 2000. Washington,DC.USA. 2001
7. J. Alfaro. Background Document. Basic Environmental Sanitation Policy. IADB. Washington, DC. USA. 1984
8. PAHO. Five Handbooks for Project's Preparation. Water and Wastewater Sector. Washington, DC. USA 1988
9. J. Alfaro. Standards for Design of Water and Wastewater Projects. Ministry of Urban Development. Peru. 1963.
10. R. Corbitt. Standard Handbook of Environmental Engineering. McGraw-Hill Publishing Co. USA. 1989
11. L. Canter.Environmental Impact of Water Resources Projects. Lewis Publishers. USA. 1985.
12. J. Peres. Nuevas tecnologias para el Tratamiento de Agua en Latinoamerica. CEPIS. Peru. 1978

13. CEPIS. Nuevos Metodos para el Diseño de Plantas de Tratamiento de Agua. Simposio. Paraguay. Peru. 1972

14. J. Peres. Plantas Modulares para el Tratamiento de Agua en Latinoamerica. CEPIS. Peru. 1982

15. J. Arboleda. Experiencias en Diseño y Construccion de Plantas de Tratamiento de Agua en Latinoamerica. CEPIS. Peru. 1983

16. J. Alfaro. Tecnologias Costo-Eficientes para el Tratamiento de Agua. Seminario. Cuernavaca. Mexico. 1993

17. D. Okun,C.Schulz. Surface Water Treatment for Communities in Developing Countries. J.Wiley & Sons. USA 1984.

18. E. Martin, T. Martin.Technologies for Small Water and Wastewater Systems. V. Nostrand R. USA. 1991

19. J. Montgomery. Water Treatment Principles and Design. J.Wiley & Sons.USA. 1985

20. J. Salvato. Environmental and Sanitation. J. Wiley & Sons. USA.1972

21. H. Hudson. Water Clarification Processes. V. Nostrand R. USA. 1981

22. G. Culp, R. Gulp. New Concepts in Water Purification. V. Nostrand R. USA. 1974

23. ASCE, AWWA,CSSE. Water Treatment Plant Design.USA.1997

24. J. Azevedo, C. Ritcher. Tratamiento de Agua. Editora Blucher. Brazil. 1991

25. J. Azevado.Innovative and Low Cost Technologies for Sewerage Systema. Brazil. 1989

26. J. Alfaro. Environmental Impact in Water and Wastewater Projects. IADB. Washington, DC. USA. 1988

27. J. Alfaro. Task Force Member. Design of Municipal Wastewater Treatment Plants. WEF. Virginia. USA. 1998

28. G. Fair, J. Geyer, D.Okun. Water and Wastewater Engineering. John Wiley and Sons. USA. 1966

29. Metcalf & Eddy. Wastewater Engineering, Treatment, Disposal and Reuse. McGraw Hill. USA.1979

30. W. Castagnino. Polucion de Agua. Modelos y Control. CEPIS. Lima, Peru. 1986

31. A. James. An introduction to Water Quality Modelling. J. Wiley & Sons. USA. 1993
32. Encibra,Stanley,CEP, Omni.Panama's Wastewater Master Plan. São Paulo. Brazil. 2002
33. J. Vaughan, C. Russell, C. Clark, D. Rodriguez, A. Darling. Investing in Water Quality. IADB. USA.2001
34. IADB. A review of the use of CVM in project analysis. Washington, DC. USA. 1998
35. WB. Better Water and Sanitation for the Urban Poor. Washington, DC. USA. 2003
36. Hernando de Soto. The Mystery of Capital. Basic Books. USA. 2000
37. Alexander Orwin.Survey on the Situation of Privatization of Water and Wastewater Utilities in several countries Environmental Probe. Canada. 1999
38. J. Alfaro. The Water Business in Latin America. A New Approach? Water Engineering and Management. USA 1987
39. G. Roth. The Role of the Private Sector in Providing Water in Developing Countries. UN. N.Y. USA. 1985
40. C. Kessides. Institutional Options for the Provision of Infrastructure. WB. Washington, DC. USA. 1993
41. Water Engineering & Management. Organization of Water Supply and Drainage Services in France. USA. 1994
42. E. Idelovitch, K. Ringskog. Private Sector Participation in Water and Wastewater in Latin America. WB. Washington, DC. USA. 1995
43. D. Haarmeyer, A. Mody. Private Capital in Water and Sanitation. WB. Washington, DC. USA. 1997
44. Water World. Handbook of Public-Private Partenership. USA. 2003
45. D. Salter.Tapping the Market. Private Sector Engagement in Rural Water Supply in the Mekong Region.Vietnam WB. Washington, DC. USA. 2004
46. IADB. Management of Water Supply and Sanitation. Case for Latin America Enterprises. Washington, DC USA. 1992
47. C. Brocklehurst, J. Janssens. Innovative Contracts, Sound Relationships:

Urban Water Sector Reform in Senegal.WB. Washington, DC. USA. 2004

48. M.Van Ginneken, R.Tylor, D. Tag. Can the Principles of Franchising be used to Improve Water Supply and Sanitation Services? A Preliminary Analysis. WB. Washington, DC. USA. 2004

Index

A

B

C

D

E

F

N

O

P

R

S

T

U

Printed in the United States
43979LVS00006BA/340-366

9 780977 263707